Heritage Building Conservation

This book provides a holistic perspective on the sustainable conservation of heritage buildings through outlining the factors that influence the preservation, operational performance and maintainability of heritage buildings and the application of new methodologies and technologies.

Using real cases from Egypt, a country which comprises a vast number of unique heritage structures, each of which is deteriorating at its own pace, this book presents a systematic, data-based approach to manage aging and deteriorating heritage assets in a cost-effective way. The authors initially provide an overview and history of conservation and maintenance work as well as the current codes and standards that regulate the preservation of these buildings. Further chapters then cover:

- The technology used to digitally document heritage buildings, including LIDAR, photo-grammetry, Heritage Building Information Modelling (HBIM), and Virtual Reality (VR) technologies
- Introducing a Maintainability Index of Heritage Buildings (MIHB) to support the decision-making and prioritization process for the maintenance of heritage buildings
- The adaptive reuse of heritage buildings
- Modelling embodied and operational energy performance
- Using Chatbot and Blockchain technology to support the management and preservation of heritage buildings

Ultimately, this book presents a useful tool for use in heritage management and highlights how the reusability of heritage buildings is critical to the creation and survival of sustainable communities. It will be useful reading for researchers, architects, engineers and especially those involved in the management of heritage buildings.

Mohamed Marzouk is Professor of Construction Engineering and Management in the Structural Engineering Department, Faculty of Engineering, Cairo University. Dr. Marzouk has more than 25 years of experience in the Civil Engineering. His experience covers different phases of projects including design, construction, monitoring, research, consulting, and project management. He also served as a consultant for many mega-construction projects in Egypt. Dr. Marzouk is acting as a Chair for Egyptian Code of Building Information Modeling. Dr. Marzouk authored and co-authored over 300 scientific publications. He is recognized and named among the top 2% of the world's most impactful scientists by Stanford University and Elsevier for three successive years 2020, 2021, and 2022.

Dr. Marzouk supervised close to 120 graduate students including 31 at the PhD level. He has acted as a Principle Investigator (PI) for eight research grants. His research interests include sustainable civil and resilient infrastructure, sustainability and green buildings, computer simulation, optimization, integrated project delivery exploring related technology including building information modeling (BIM), 3D laser scanning, virtual/augment reality, digital twins, and artificial intelligence. Dr. Marzouk has been involved in several academic and scientific committees. He has been awarded several excellence awards and scholarships.

Heritage Building Conservation

Sustainable and Digital Modelling

Edited by
Mohamed Marzouk

Routledge
Taylor & Francis Group

LONDON AND NEW YORK

First published 2023
by Routledge
4 Park Square, Milton Park, Abingdon, Oxon OX14 4RN

and by Routledge
605 Third Avenue, New York, NY 10158

Routledge is an imprint of the Taylor & Francis Group, an informa business

British Library Cataloguing-in-Publication Data
A catalogue record for this book is available from the British Library

Library of Congress Cataloging-in-Publication Data
Names: Marzouk, Mohamed, author. | El Sharkawy, Maryam, author. |
Mohamed, Hoda, author. | Metawie, Mohamed, author.
Title: Heritage building conservation : sustainable and digital modelling /
Mohamed Marzouk, Maryam El Sharkawy, Hoda Mohamed,
and Mohamed Metawie.
Description: Abingdon, Oxon ; New York, NY : Routledge, [2023] | Includes
bibliographical references and index. Identifiers: LCCN 2023004794 |
Subjects: LCSH: Historic buildings—Conservation and restoration.
Classification: LCC TH3301 .M27 2023 | DDC 363.6/9—dc23/
eng/20230215
LC record available at https://lccn.loc.gov/2023004794

ISBN: 978-1-032-37492-5 (hbk)
ISBN: 978-1-032-41317-4 (pbk)
ISBN: 978-1-003-35748-3 (ebk)

DOI: 10.1201/9781003357483

Typeset in Times New Roman
by codeMantra

Contents

Preface

Having a strong cultural heritage is crucial to a nation's long-term prosperity because it serves as a constant resource for maintaining national identity and historical memory. Heritage buildings, which are old and historically significant, play
an important part in modern life. Those buildings are considered significant de-
piction of nations' records and history. However, these buildings are constantly
undergoing a variety of degradation processes, including aging, weathering, and
use-related damage. Accordingly, sustainable heritage conservation has played an
important role in preserving, repairing, documenting, and reusing these heritage
buildings while maintaining their significant values. Heritage building's conservation involves different levels of intervention, such as preservation, restoration, and
adaptive reuse depending on the deterioration rate of the building, its functionality, and its cultural significance. Moreover, digital documentation tools (such as
3D computer graphics, photogrammetry, and laser scanning), energy assessment
tools, virtual twins, and other innovations helped to improve heritage buildings'
documentation, conservation, restoration, and rehabilitation in the digital era. The
procedure begins with appropriate digital recording and damage assessment in heritage structures, identifying the factors that affect the operational performance of
the building with respect to its energy performance, thermal comfort, and daylighting performance; subsequently, ideal scenarios could be recommended based on
the history and functionality of the building. Over the past few years, both the
BIM industry and the heritage authorities or communities have become increasingly aware of heritage building information modeling (HBIM). The sustainable
heritage paradigm has undergone a significant shift, as evidenced by the expanding
processes and tools of HBIM.

The main purpose of this book is to provide a holistic perspective on the sustainable conservation of heritage buildings by outlining the factors that influence the
preservation, operational performance and maintainability of heritage buildings,
and the application of new methodologies and technologies. This book is divided
into seven chapters that demonstrate its main purpose. This is accomplished by
discussing the various sustainable conservation intervention levels and the sustainable conservation shift in the digital era in Chapter 1. Following that, as in Chapter
2, previous research efforts in the field of sustainable heritage buildings are presented through a science mapping analysis for the past two decades, highlighting

the important areas for future research in this field. Chapter 3 identifies the parameters that have an impact on the maintenance and preservation of heritage buildings in the Egyptian context, as well as introducing the Maintainability Index of Heritage Buildings (MIHB), which captures the current status of these buildings and their severity condition. In addition, the metrics and techniques used to evaluate the viability of heritage conservation and reusability are highlighted in Chapter 4. To demonstrate the efficacy of modern digital documentation technologies for the preservation of historic buildings, Chapter 5 investigates the application of LIDAR and HBIM technologies within a framework adaptable to Egyptian Heritage Documentation activities. Furthermore, Chapter 6 discusses the factors that influence the operational performance of heritage buildings, as well as providing a framework for analyzing and assessing the operational performance of heritage buildings and proposing sustainable intervention strategies to improve the building's performance. Finally, in Chapter 7, a data-driven Chatbot framework using Blockchain technology is proposed to aid in the management of maintenance tasks using HBIM.

Contributors

Hoda Abdelrazik is Head of Sector, Ministry of Housing Utilities and Urban Communities, Cairo, Egypt.

Maryam El-Maraghy is Researcher, Construction Engineering Technology Lab, Faculty of Engineering, Cairo University, Giza, Egypt.

Maryam El Sharkawy attained her Ph.D. from Cairo University. She has worked as a part-time Lecturer Assistant and Research Assistant at the Faculty of Engineering (Cairo University) and the School of Science and Engineering (American University in Cairo). From 2012 to -2016, she worked as an Architect for ECG Engineering Consultants Group S.A.

Nouran Labib is an MSc Student, Integrated Engineering Design Management Program, Faculty of Engineering, Cairo University, Giza, Egypt.

Mohamed Marzouk is Professor of Construction Engineering and Management, Structural Engineering Department, Faculty of Engineering, Cairo University, Giza, Egypt.

Mahmoud Metawie is Assistant Professor in the Structural Engineering Department, Faculty of Engineering, Cairo University, Egypt.

Basma Mohamed is Assistant Lecturer, Structural Engineering Department, Faculty of Engineering, Cairo University, Giza, Egypt.

1   Introduction

Mohamed Marzouk

Professor of Construction Engineering and Management, Structural Engineering Department, Faculty of Engineering, Cairo University, Postal code 12613, Giza, Egypt.

Maryam El-Maraghy

Researcher, Construction Engineering Technology Lab, Faculty of Engineering, Cairo University, Postal Code 12613, Giza, Egypt

Maryam ElSharkawy

Instructor, Faculty of Engineering, Cairo University, Postal code 12613, Giza, Egypt

1.1   General

The world's heritage today is considered a huge value to humanity. An existing material hub and carbon bank within the built environment represent a significant matter of civilization history. Past is gold; however, the civilization of history is at risk when it comes to the pollution crisis and high energy consumption levels in cities. Fortuitously, sustainable heritage in the digital era carries multiple capabilities and potentials to boost economic, environmental, and social values. Besides, the rise of multiple international interventions and policies to increase awareness has motivated a sustainable paradigm. In the Digital era, digital documentation tools, damage & energy assessment tools, virtual twins, facility management, and more succeeded in progressing heritage documentation, conservation, restoration, and rehabilitation. The process may begin with the suitable means of digital documentation and assessment of damage in heritage buildings; then, optimum scenarios could be suggested according to the legend of each unique piece of art.

1.2   Heritage Buildings: An Overview

Heritage is a continuous resource that promotes identity and memory and is essential to attaining sustainable development of nations [1, 2]. Buildings, structures, artifacts, and locations that are historically, artistically, and architecturally significant are part of the heritage. A city's cultural and natural resources, and tangible and intangible elements could be considered heritage. Thus, it promotes socio-economic regeneration, strengthens social well-being, reduces poverty, increases regional appeal, and increases long-term tourism advantages [1, 2]. According to UNESCO, the world heritage can be categorized into natural heritage (natural

DOI: 10.1201/9781003357483-1

sites, natural features, and geological and physiographical formations) and cultural heritage (monuments, groups of buildings, and sites) [3, 4]. Cultural heritage is divided into two categories: tangible and intangible cultural heritage. Tangible cultural heritage is considered the remaining assets that are resulted from past human activities and development; they have a significant value that reflects the past culture, developments, and achievements. Tangible heritage is divided into movable and immovable heritage [5, 6]. The Moveable heritage items include coins, sculptures, paintings, papers, and furniture. Whereas, Immovable heritage assets include heritage buildings, archaeological sites, monuments, and undersea heritage such as shipwrecks, underwater remains, and cities [5, 6]. On the other hand, intangible cultural heritage as defined by UNESCO is the transmitted knowledge, skills, expressions, and practices between different generations of a group or community that are created according to their history, identity, and cultural diversity [7]. According to UNESCO [7], they are divided into oral traditions and expression, performing arts, social practices, rituals and festive events, traditional craftsmanship, and knowledge and practices concerning nature and the universe. Figure 1.1 shows the UNESCO heritage classification [4–7].

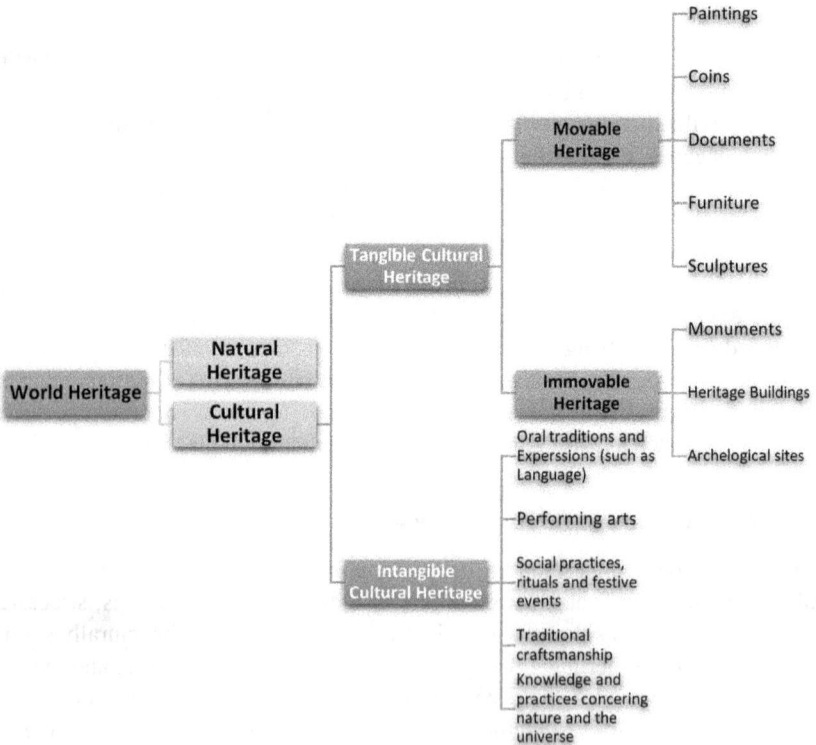

Figure 1.1 Heritage Classification (Adopted from [4])

Historical structures known as heritage buildings have a significant role in contemporary society [1]. Heritage structures are vital symbols of people's history and culture because they are physical reminders of their past. Community identity and stability can be supported by the presence of these buildings [8]. Heritage buildings can be divided into three groups: (1) monuments and buildings of great architectural interest; (2) buildings erected before a certain historical time; and (3) buildings that reveal exceptional structural and technological components [9].

The values attached to a heritage property are no longer viewed as an unchanging constant but rather as something that changes over time and space [10]. Values attached to a place may alter when demographics and technology affect the requirements of one stakeholder group. One generation's values on a location may differ from those of preceding generations. Complicating matters even further is the possibility that different groups of interested parties may place varying values on the same location, some of which may be in direct conflict with one another. According to Araoz [10], there are numerous and obvious indications that a new heritage perspective has evolved in dealing with heritage buildings, such as (1) the absence of actual fabric to sustain official cultural identification as a heritage, (2) control of social processes deemed essential to the place's relevance, (3) usage of cultural sites as a solution to mitigate poverty by development agencies, (4) acceptance of duplicate reconstructions as valid replacements for originals, (5) excessive replacement-in-restoration, (6) analysis of archeological sites to improve the attractiveness and cultural accessibility, and (7) the constantly growing tourism infrastructure that degrades the heritage environment.

The continuous enhancements in the environmental, economic, and social conditions in global cities, buildings, and environments are becoming our common future rule following sustainability norms. The main aim is to protect our natural resources and ensure their availability for future generations. Thus, the process cannot occur with artifacts and heritage negligence or failure to preserve. This links the idea of sustainability to the idea of economic growth that doesn't compromise environmental safeguards [11].

This means that conserving historical buildings helps achieve the three sustainability pillars. Historic buildings should be preserved instead of demolished because they provide insight into past societies and encounter huge material hubs; building preservation and adaptive reuse that considers a building's age, size, and potential can teach history to future generations [12]. According to Pintossi [13], conservation's function has moved from preservation to inclusion in a larger urban regeneration strategy and sustainability, necessitating widespread engagement and the use of multiple academic disciplines. Consequently, sustainable development is aided by protecting material and immaterial cultural heritage. The preservation of cultural heritage requires knowledge of not only the building's physical characteristics and the processes that cause deterioration but also the exterior social, political, economic, and cultural factors that impact the city's growth and development [14]. Defining the social, economic, and environmental performance indicators must be considered when preserving built heritage as a sustainable practice and developing public policies to support it. To be able to promote sustainability, an agenda must include historic preservation as well as environmental issues. Many historic and culturally significant buildings are being reused rather than demolished as part of a broader rehabilitation strategy to promote sustainability in the built environment [15].

1.3   A Sustainable Heritage Paradigm

Various forms of deterioration, such as aging, weathering, and use-related deterioration, occur continuously in heritage buildings [16]. The heritage conservation community was based on the basis that all place's value was derived from its physical signs. For more than two centuries, conservationists' theories and practices have evolved to keep a place's form and space as stable as possible [10]. The degree of building deterioration depends on the building's structure, fabrics, and repairs; as a result, repair techniques may vary across diverse building cultures and technologies. However, the primary objective of repairing and preserving a heritage building is to keep and preserve the original architect's work for the present and future generations [16].

Heritage buildings conservation involves the actions taken to preserve a site's historical and cultural significance. This category includes maintenance, which may also contain preservation, and restoration, depending on the nature of the issue at hand and the available resources. Conservation is the process of preventing deterioration of a location's historical and cultural significance [16]. Depending on their historical significance, functioning, current condition, or obligatory code requirement, different conservation methods, such as preservation, rehabilitation, or adaptation to new functions, can be applied to buildings that no longer serve their original purpose [17]. All of these interventions on historic structures necessitate some change to preserve the building and its assets for future generations while also making it usable in the present [16]. As stated by Zhang and Dong [8], the first step in heritage building conservation is to analyze the level of conservation interventions that would take place in order to determine the minimal intervention. Different intervention practices will affect the outcome and are influenced by the specific constraints of each site. Intervention should be as minimum as possible and proportional to the projects' safety measures to guarantee the least disruption of a site's or structure's historical value while ensuring appropriate safety and durability measures [8].

According to international charters' definitions, the various degrees of building conservation intervention include prevention, maintenance, repair, renewal, reuse, and new design [8]. The optimum level of intervention starts with the prevention of deterioration [18], followed by maintenance is considered the lowest level of intervention when it comes to any building's alteration, moving to the new design is considered the highest level of intervention [8], the definition for these terms is explained below as follows [8, 18]:

- Prevention of deterioration [18].
- Maintenance: Trying to maintain a place's current appearance by delaying degradation without affecting its functionality, such as preservation and refurbishment.

- Repair: Replacing older materials with new ones, such as renovation, restoration, and consolidation.
- Renewal: the act of reviving or revitalizing something.
- Reuse: adaptive reuse or appropriate uses, in which the role of architectural conservation shifts from protection to urban development and sustainability, as well as the incorporation of new functions to maintain the suitability of heritage buildings, such as adaptive reuse, transformation, compatible use, and rehabilitation, and
- New Design: a new design concept meant to reduce the negative effects of urban infill on historic surroundings with the least amount of interference, such as regeneration and reconstruction [8].

Figure 1.2 shows the levels of heritage building conservation intervention components, which are described below in detail.

Preservation is "the act or process of implementing the necessary measures to maintain the existing form, integrity, and materials of a historic property" [16]. Keeping a building in good condition and limiting the amount of structural damage it sustains is called preservation, which is related to maintenance. It involves reducing the rate of deterioration of a building's structure to preserve its current state, utilizing appropriate repair methods to retard the deterioration of an existing building, and attempting to avoid the high rate of decay caused by nature [16, 19]. Indirect methods are also included, such as the monitoring process, which involves

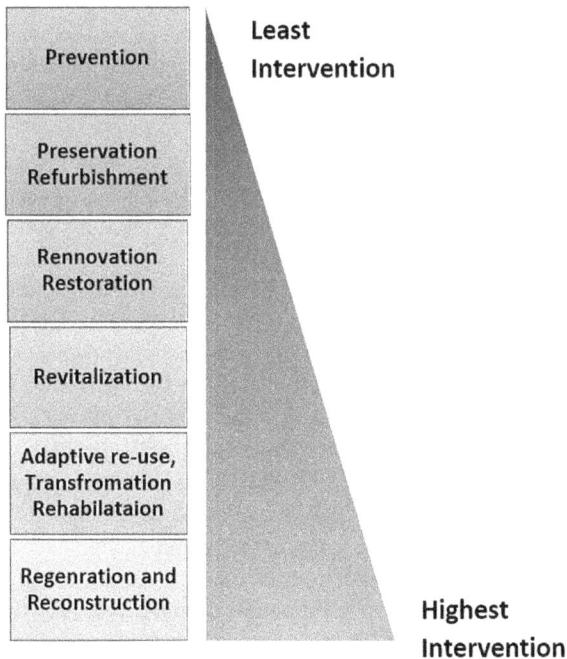

Figure 1.2 Degrees of Building Conservation Interventions (Adopted from [8])

the continuous measurement, surveying, and evaluation of changes in material attributes and environmental conditions. Doing so preserves the buildings or their heritage value and keeps them in good condition by recognizing the structure or part as a physical record of its time, place, and function [20].

Refurbishment is the improvements made to an existing structure, such as improvements to the building's envelope in terms of both form and function and major maintenance and repair work to bring the building as stated by the current standards. It can be refurbished to preserve a historic building's architectural integrity if the work is done with appropriate materials and methods [20].

Renovation refers to bringing outdated features, systems, and fixtures up to code, which may include increasing the building's energy efficiency. In most cases, renovations meant to bring a historic building up to modern standards of comfort are not conservation actions since they do not adequately preserve the building's significance. When modern materials and technical installation are used, it is impossible to bring back the original materials, finishes, character-defining features, and energy performance [20].

Restoration means to restore the physical and/or aesthetic condition of an old building to that of a certain date or event. Restoring an existing building to its as-previously-known state does not necessitate using any new building materials. Restoration of heritage buildings entails restoring a structure to its former form using its original materials and new technologies [16]. The scope of such projects is typically limited to extensive renovations of abandoned or damaged structures. Large-scale renovations to the building or its components are possible [20].

Consolidation considers alterations of the decayed building elements such as fabric, electrical, and mechanical systems, but no new additions are allowed [8, 18].

Revitalization can refer to cleaning up a building or conservation area, which involves any action taken to improve the environment and infrastructure of a historic area or building. An approach to sustainable design that considers the behavioral change, requirements, and patterns of the surrounding communities is to provide a wide range of new functionalities [21]. It includes revitalizing the area around historic structures while addressing pressing issues, including access to running water, sewage, and power, and strengthening the buildings themselves [8].

Adaptive Reuse involves modifications to a building (other than regular maintenance) that enhance its capability, efficiency, and effectiveness, preparing the place for a new use that is compatible with the existing one [16]. Adaptive reuse of a building can be applied only when the adaptation has a negligible impact on the cultural value of the place. The best way to keep a historic structure standing is to adapt it for a new purpose [16]. Adaptive reuse of heritage buildings for new purposes can assist communities, governments, and developers decrease urban expansion's environmental and social costs. Reusing heritage buildings can help transform a region while making them accessible and usable to the public [15]. Deconstruction and reconstruction are unnecessary when using adaptive reuse. Reuse is critical to sustainable development because of its positive impact on the

environment and the economic, social, and cultural benefits of preserving a historically significant building [22].

One option to reduce carbon emissions is incorporating environmental design into
the adaptive reuse of historic buildings [22]. Adaptive reuse is a great tool for repur-
posing underutilized buildings into community assets, saving money on land and
construction, reviving communities, and limiting overdevelopment [17]. Designers
face substantial problems when tasked with implementing adaptive reuse. Choosing
the best suitable use for the building within its environment is critical to protect the
heritage site's cultural significance [12]. There are several factors to consider while
planning for the adaptive reuse of a historic property, including the structure's historical significance, physical features, and future uses [12]. Internal and external factors
have been identified as the main drivers of adaptation. Technological advancement,
sustainability, economic expansion, and the spread of urbanization are all global
challenges attributed to external forces. Instead, internal factors are those that origi-
nate from within the organization itself. There are many causes of a building's quick
decline in use, including neglect and abandonment [17]. The most important issue
with adaptive reuse is the lack of thorough investigation before deciding the new role
for historic buildings. For an adaptive reuse project to be successful, a heritage building needs to be financially, socially, and environmentally viable [12].

Transformation refers to introducing new uses and significance to old buildings and locations. In a historical context, modernizing the building's function by
altering its function [8]. The goal of the building transformation is to restore a
building's usefulness and ensure its continued existence [21].

According to Soewarno et al. [21], there are three categories of the transformation process, which are:

1   Core element: to gradually transform changing building elements with keeping
    identical features.
2   Peripheral element: to transform less significant building elements and track
    how easily they can be modified or substituted.
3   New element transformation: The change as the owner adapts to a new culture.
    It is integrated into the architecture.

Compatible use is having no effect on, or just a little effect on, a place's cultural
significance. If a building or site has cultural significance, it should be used in a
way that doesn't compromise its original purpose [8, 19].

Rehabilitation is converting a property to a usage compatible with its current
purpose. Some extensions and even large structural changes are possible components of upgrading [20]. Rehabilitating a building's architecture entails bringing
it up to date with construction standards, energy efficiency rules, and comfort and
usage guidelines [23]. The goal of rehabilitating a historic structure is to find a
modern use complementary to the building's unique history and architecture. It
must convey the property's worth by making a few alterations to the building's
original structure, fixtures, and finishes [20]. Building rehabilitation assumes a new

stage in an existing building's life cycle. It involves mostly reusing the components and structures that have previously been erected, with only minor material and energy additions [23]. When a building is rehabilitated, it provides an opportunity to improve its environmental performance and lower the energy it uses. In contrast to demolishing and building new structures, rehabilitation provides environmental, social, and economic benefits. When paired with preservation measures for cultural or historical significance structures, architectural rehabilitation preserves and promotes a crucial social capital: built heritage [23].

Regeneration is a planning approach that aims to revitalize, improve, and incorporate the existing historic environment into a new, contemporary, supporting initiative [24]. This could be achieved by redesigning or decorating the current heritage buildings [8].

Reconstruction is considering adding new materials and replacing degraded textiles in cases where the building's state has been largely or completely lost, and reconstruction will return it to an earlier state [8, 18, 19].

1.4   Heritage in the Digital Era

Heritage buildings, archaeological sites, and cultural locations need to be documented for their identification, protection, and preservation to be given the attention they deserve [14]. Built-heritage documentation requires the use of 3D geometry modeling and the management of semantic knowledge stored in databases. 3D computer graphics, photogrammetry, laser scanning, Geographic Information System (GIS), ontologies, and the more contemporary Building Information Modeling (BIM) advancements are only some of the information tools now used for architectural heritage modeling, management, and conservation [25].

Previously, computer-aided design (CAD) allowed professionals to develop and implement their ideas without established methods to communicate and retain the accompanying metadata. Nowadays, by using BIM technology, "smart" models may be created with all the relevant data needed for construction projects included in a single, streamlined file [26]. The term BIM is more suited to describe a method than a software program [26]. BIM, or Building Information Modeling, is a relatively new approach that merges 3D modeling with data management [25]. It's a digital model of the project in three dimensions (3D), with information flowing between each component.

In contrast to CAD, which is simply a 3D model that enables the creation of plans, sections, etc., but which is still composed of purely geometrical parts, BIM enables the creation of smart models with all their elements linked to a database containing all the relevant metadata. [26]. The latest generation of BIM technology makes it possible to combine various documentation data into a single, comprehensive building model for the first time [27]. As a result of the growing popularity of BIM, a new field has emerged: historic/heritage BIM (HBIM). HBIM is an expansion of BIM used to manage data and create 3D models of historic buildings [14, 25]. The HBIM system incorporates parametric databases with 3D visualization. As a result, visual and numerical data flows can be managed together [28].

According to Yang et al. [25], HBIM supports (i) the building's history through a comprehensive survey and parametric modeling of its geometry; (ii) the building's sub-elements through information on their attributes, materials, and relationships; and (iii) the building's potential deformations and changes over time [25].

To recreate a historical building using existing description data (such as historical records, bibliographic references, pictures, and sketches), HBIM is well suited, particularly for lost or vanished cultural properties [25]. The monitoring of heritage building conservation, refurbishment, and retrofitting is aided by HBIM, which helps predict how it will perform in the long term [27, 29]. Methodologies and techniques have been developed to manage and integrate data for the documentation of built heritage, specifically the 3D representation of buildings using computer graphics and digital acquisition techniques [14]. According to Khalil et al. [27], there are two stages in the digital documentation of heritage buildings. The first stage is gathering all essential data utilizing various data-gathering methods, such as surveying and raw monitoring data. The second stage is the data interpretation stage, in which the surveyed raw data of a heritage building is converted into valuable information [27]. Various digital documentation procedures have recently been implemented, including laser scanning and photogrammetry for documenting the physical structure of buildings and defining texture, as well as the use of nondestructive techniques for capturing current conditions and deterioration, such as thermal imaging and sensors. These techniques have grown in popularity, particularly for creating textured 3D models that depict the structure of the heritage building and its materials [30]. Digital photogrammetry, in conjunction with 3D laser scanning, is increasingly used for heritage building documentation [26, 31]. They build a digital model from recorded data that is virtually equal to the physical geometry of the area. Digital documenting methods have been acknowledged as vital in conservation rather than active restoration [31]. Where digital photogrammetry images are recorded from multiple perspectives, the process is based on the triangulation principle. When combined with laser scanning, high-resolution photographs of material textures and data on material deterioration can be obtained [26]. With photogrammetry and 3D laser scanning, it is now possible to produce a point cloud that represents the complete geometry of a building [32]. Identifying building components and degradation is important, especially for conservation intervention strategies. In the domains of geometric documentation and architectural documentation, the use of laser scanning and other image-based technologies is routine practice. These huge amounts of data can help protect cultural heritage resources and select appropriate conservation strategies [30]. Some terrestrial laser scanners (TLS) use the triangulation method as well. The scanning procedure produces a point cloud of the scanned item, which is then turned into a 3D model. TLS can be linked to a GIS to evaluate geographical data and locate the scanned object using Cartesian coordinates [26].

HBIM awareness has been increasing rapidly in the past few years in both the BIM industry and the heritage authorities or communities. HBIM evolving processes and tools revealed a major transformation in the sustainable heritage paradigm. Imagine the added value of a digital replica, including physical and nonphysical information for a sophisticated artifact building to the heritage communities, regardless of the uncountable sorts of analysis and assessments of the building via digital tools. These all pour into aiming at heritage sustainability.

A wide range of technological advances has been incorporated to facilitate the process of HBIM, including physical and nonphysical data retrieval. However, Counsell and Taylor [33] set the requirements for the HBIM framework, which is essential to support the full range of cultural heritage environments.

1.5   Conclusion

The identity of nations and the most precious asset any city owns is mainly represented through heritage. Thus, it is considered a global drive toward preserving heritage worldwide. The various means and degrees of heritage conservation have been identified, and some of the latest technologies are involved in the conservation processes. Utilizing future and current technologies to preserve the past is essential to preserve historical elements. Sustainable heritage is shifting in a digital era. With the newly emerging tools, some are even dedicated and specialized for heritage; we are currently more capable of preserving them for all future generations. Even the effect of time on heritage can be documented in various means to effectively deliver the originals and the complete story to those who are not present. Further, the global policies and regulations aiming at heritage conservations are to be demonstrated, along with HBIM current advancements.

References

[1] Al-Sakkaf, A., T. Zayed, and A. Bagchi, A Review of Definition and Classification of Heritage Buildings and Framework for their Evaluation. in *2nd International Conference on New Horizons in Green Civil Engineering (NHICE-02).* 2020. Victoria, BC.

[2] ICOMOS, The Paris Declaration. On Heritage as a Driver of Development. 2011. ICOMOS, Paris.

[3] UNESCO, Convention Concerning the Protection of the World Cultural and Natural Heritage. 1972. Paris.

[4] Petti, L., C. Trillo, and B.N. Makore, Cultural heritage and sustainable development targets: A possible harmonisation? Insights from the European perspective. *Sustainability*, 2020. **12**(3), 926. DOI: 10.3390/su12030926.

[5] Cosovic, M., A. Amelio, and E. Junuz, Classification Methods in Cultural Heritage. in *CEUR Workshop Proceedings* (Vol. 2320, pp. 13–24). 2019. CEUR-WS, Pisa.

[6] Kurniawan, H., et al., E-Cultural Heritage and Natural History Framework: An Integrated Approach To Digital Preservation. in International Conference on Telecommunication Technology and Applications (IACSIT). 2011. Singapore.

[7] UNESCO, Basic text of 2003 Convention for the Safeguarding of the Intangible Cultural Heritage, UNESCO,2022, Paris.

[8] Zhang, Y. and W. Dong, Determining minimum intervention in the preservation of heritage buildings. *International Journal of Architectural Heritage*, 2021. **15**(5): p. 698–712.

[9] Cabeza, L.F., A. de Gracia, and A.L. Pisello, Integration of renewable technologies in historical and heritage buildings: A review. *Energy and Buildings*, 2018. **177**: p. 96–111.

[10] Araoz, G.F., Preserving heritage places under a new paradigm. *Journal of Cultural Heritage Management and Sustainable Development*, 2011. **1**(1): p. 55–60.

[11] Nocca, F., The Role of Cultural Heritage in Sustainable Development: Multidimensional Indicators as Decision-Making Tool. *Sustainability*, 2017. **9**(10), 1882.

[12] Mısırlısoy, D. and K. Günçe, Adaptive reuse strategies for heritage buildings: A holistic approach. *Sustainable Cities and Society*, 2016. **26**: p. 91–98.

[13] Pintossi, N., D. Ikiz Kaya, and A. Pereira Roders, Assessing cultural heritage adaptive reuse practices: Multi-scale challenges and solutions in Rijeka. *Sustainability*, 2021. **13**(7), 3603. DOI: 10.3390/su13073603.

[14] Ronzino, P., A. Toth, and B. Falcidieno, Documenting the structure and adaptive reuse of Roman amphitheatres through the CIDOC CRMba Model. *Journal on Computing and Cultural Heritage*, 2022. **15**(2): p. 1–23.

[15] Bullen, P.A. and P.E.D. Love, Adaptive reuse of heritage buildings. *Structural Survey*, 2011. **29**(5): p. 411–421.

[16] Mehr, S.Y., Analysis of 19th and 20th century conservation key theories in relation to contemporary adaptive reuse of heritage buildings. *Heritage*, 2019. **2**(1): p. 920–937.

[17] Elsorady, D.A., Assessment of the compatibility of new uses for heritage buildings: The example of Alexandria National Museum, Alexandria, Egypt. *Journal of Cultural Heritage*, 2014. **15**(5): p. 511–521.

[18] Khalil, A.M.R., N.Y. Hammouda, and K.F. El-Deeb, Impslementing sustainability in retrofitting heritage buildings. Case study: Villa Antoniadis, Alexandria, Egypt. *Heritage*, 2018. **1**: p. 57–87. DOI: 10.3390/heritage1010006.

[19] Australia ICOMOS, The Burra Charter: The Australia ICOMOS Charter for Places of Cultural Significance. 2013. ICOMOS, Burwood.

[20] Bertolin, C. and A. Loli, Sustainable interventions in historic buildings: A developing decision making tool. *Journal of Cultural Heritage*, 2018. **34**: p. 291–302.

[21] Soewarno, N., T. Hidjaz, and E. Virdianti, The sustainability of heritage buildings: Revitalization of buildings in the Bandung Conservation Area, Indonesia. *WIT Transactions on Ecology the Environment*, 2018. **217**: p. 687–697.

[22] Yung, E.H.K. and E.H.W. Chan, Implementation challenges to the adaptive reuse of heritage buildings: Towards the goals of sustainable, low carbon cities. *Habitat International*, 2012. **36**(3): p. 352–361.

[23] Munarim, U. and E. Ghisi, Environmental feasibility of heritage buildings rehabilitation. *Renewable and Sustainable Energy Reviews,* 2016. **58**: p. 235–249.

[24] Said, S.Y., et al., Sustaining old historic cities through heritage-led regeneration. *WIT Transactions on Ecology the Environment*, 2013. **179**: p. 12.

[25] Yang, X., et al., Review of built heritage modelling: Integration of HBIM and other information techniques. *Journal of Cultural Heritage*, 2020. **46**: p. 350–360.

[26] Pocobelli, D.P., et al., BIM for heritage science: A review. *Heritage Science*, 2018. **6**(1): p. 30.

[27] Khalil, A., S. Stravoravdis, and D. Backes, Categorisation of building data in the digital documentation of heritage buildings. *Applied Geomatics*, 2021. **13**(1): p. 29–54.

[28] Khalil, A. and S. Stravoravdis, H-BIM and the domains of data investigations of heritage buildings current state of the art. *International Archives of the Photogrammetry, Remote Sensing and Spatial Information Sciences*, 2019. **XLII-2/W11**: p. 661–667.

[29] Al-Sakkaf, A. and R. Ahmed, Applicability of BIM in heritage buildings: A critical review. *International Journal of Digital Innovation in the Built Environment (IJDIBE)*, 2019. **8**(2): p. 20–37.

[30] Delegou, E.T., et al., A multidisciplinary approach for historic buildings diagnosis: The case study of the Kaisariani Monastery. *Heritage*, 2019. **2**(2): p. 1211–1232.

[31] Jo, Y.H. and S. Hong, Three-dimensional digital documentation of cultural heritage site based on the convergence of terrestrial laser scanning and unmanned aerial vehicle photogrammetry. *ISPRS International Journal of Geo-Information*, 2019. **8**(2), 53.

[32] Rocha, G., et al., A scan-to-BIM methodology applied to heritage buildings. *Heritage*, 2020. **3**(1): p. 47–67.

[33] Counsell, J. and T. Taylor, What are the Goals of HBIM?, in Heritage Building Information Modelling. (Eds. Arayici, Y., J. Counsell, L. Mahdjoubi, G. Nagy, S. Hawas, and K. Dweidar), Routledge, New York, 2017, p. 15–31.

2 Science Mapping Analysis of Sustainable Heritage Buildings

Basma Mohamed

Assistant Lecturer, Structural Engineering Department, Faculty of Engineering, Cairo University, Postal Code 12613, Giza, Egypt

Mohamed Marzouk

Professor of Construction Engineering and Management, Structural Engineering Department, Faculty of Engineering, Cairo University, Postal Code 12613, Giza, Egypt

2.1   General

Nowadays, the conservation of heritage buildings and accomplishing sustainability goals are among the most important pursuits for nations. Hence, many researchers have targeted "Sustainable Heritage Buildings" (SHB) in their studies and conducted review papers about it. This study conducts a science mapping analysis of SHB throughout 2002–2022 to address the subjective limitations of previous review studies. Ninety peer-reviewed research articles were subsequently extracted from the Web of Science as a Core Collection database. The proposed research methodology results revealed some interesting findings regarding the annual publication trends, most cited articles, most active researchers, geographic distribution of the publications, leading academic journals, and the thematic landscape of the field. The study demonstrated that the important issues worthy of future exploration are those related to "Thermal Comfort, Energy Efficiency, and Energy Retrofit." Various entities can benefit from the presented results concerning conducting state-of-the-art research and addressing the practical issues of sustainably preserving heritage buildings.

2.2   Introduction

The protection and safeguarding of the world's cultural and natural heritage are one of the targets of the goals stated in the United Nations Sustainable Development Goals (SDGs) [1]. This requires preventing losses through timely interventions [2]. However, these efforts are challenged by multiple stakeholders and the many values associated with the decision-making process. Consequently, SHB attracted the attention of many researchers. For instance, Posani et al. [3] presented an overview of various envelope retrofit interventions that can be applied to historic buildings to improve their resilience and energy efficiency while conserving their architectural values. They also proposed a fast assessment method for thermal insulation

DOI: 10.1201/9781003357483-2

solutions to assess their compatibility with historical components as a first step to defining a numeric criterion for the same purpose. Cabeza et al. [4] investigated tailored energy retrofit strategies that can be applied in heritage buildings without compromising their architectural values. These strategies include energy efficiency measures, such as internal envelope insulation, cool coating, and window retrofitting, integrating renewable energies like solar energy using UV panels and solar lighting, and using heat pumps. Additionally, Pereira et al. [5] presented a multidisciplinary framework to address the challenges facing higher education heritage buildings concerning maintaining their identity, selecting optimum measures for energy retrofitting, and adapting to global warming and climate change. Nevertheless, these review studies have qualitatively interpreted the findings and results. They also lack a general dynamic quantitative analysis of the SHB research domain over time. Hence, this established the need for an objective quantitative method to review the literature about SHB.

Given the remarkable and unprecedented growth of scientific documents being published in different research areas, novel techniques are necessary to enable a comprehensive understanding of such areas by evaluating them and extracting their patterns, new trends, knowledge gaps, and promising topics for future exploration [6]. Bibliometrics is a set of quantitative methods that examine research via the article metadata provided in bibliographic databases (e.g., Scopus, Web of Science Core Collection). This metadata includes a variety of information about each publication, including its title, authors and their affiliations, keywords, abstract, and citation information. Science mapping analysis is a branch of the bibliometric analysis approaches concerned with extracting knowledge from a research domain's collaboration, citation, and conceptual networks [6]. This is accomplished by studying the main scientific actors (authors and countries), prominent academic journals (sources), and the various research field themes.

Given the great importance of sustainably preserving heritage buildings and addressing the relevant limitations in the body of knowledge, this study aims to comprehensively and quantitatively review the research literature conducted about SHB. The research methodology is divided into five steps, as per Figure 2.1.

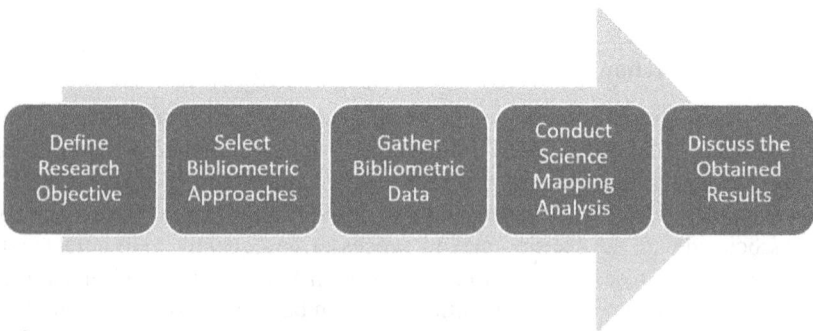

Figure 2.1 Proposed Research Methodology

Table 2.1 Bibliometric Analyses

Bibliometric Analysis	Input	Output
Coauthorship analysis	Authors & countries	Collaboration networks
Citation analysis	Journals	Citation networks
Cooccurrence, cluster, and theme analyses	Keywords	Conceptual networks (topics and themes)

2.3   Bibliometric Analyses

The current research study aims to offer entities concerned with SHB a quantitative and comprehensive understanding through bibliometric analysis and science mapping. This is accomplished by analyzing publication trends and identifying influential research studies. In addition, collaboration networks of scientific contributors are examined through coauthorship analysis. Citation analysis is employed to study the citation networks of sources (peer-reviewed journals). Regarding the conceptual structure of the research domain, it is analyzed via cooccurrence, cluster, and theme analyses. Table 2.1 illustrates the bibliometric analyses conducted and explains their required inputs and resulting outputs.

The bibliographic data of the present study was extracted from the Web of Science, which is preferred to other databases such as Scopus because it can narrow down the conducted search to include only papers related to the "Construction and Building Technology" category. The search was conducted in November 2022 with the following keywords: Keywords: ("heritage building*" OR "historic building*" OR "cultural heritage*" OR "built heritage" OR "heritage site*") AND "sustainabl*." The truncation symbol "*" replaces several letters to ensure that all the keywords' derivatives are included in the search. Furthermore, the date range was set to 2002–2022. The results were refined by selecting only peer-reviewed journal papers published in English. Thus, this resulted in 90 peer-reviewed research publications for the subsequent analysis.

2.4   Science Mapping Results

2.4.1   Annual Publication Trends

Microsoft Excel was utilized to aggregate the yearly publications to assess the academic output of SHB. Figure 2.2 illustrates the resulting publication trend. It can be concluded that the first publications were conducted around 2006, and there was only zero to one publication per year up until 2014. Starting in 2015, there was a dramatic increase to four or more publications, with a peak of 21 articles in 2021. This shows that tackling SHB is a promising research field worthy of further exploration.

2.4.2   Influential Articles

Publications about SHB are assessed for their citation scores to assess overall influence as well as their influence within the extracted bibliographic database through

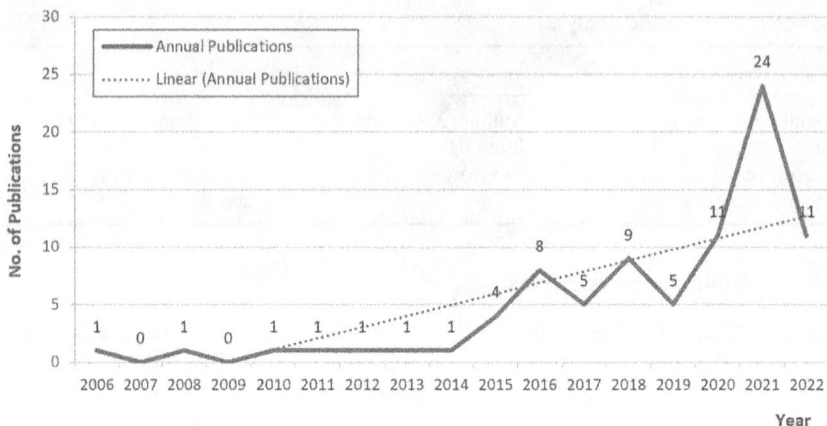

Figure 2.2 Annual Publications

Table 2.2 Most Cited Articles

Article	Title	GS	LC
Rodrıguez et al., 2016 [9]	Spatial-temporal study on the effects of urban street configurations on human thermal comfort in the world heritage city of Camagüey-Cuba	79	0
Vardopoulos, 2019 [8]	Critical sustainable development factors in the adaptive reuse of urban industrial buildings. A fuzzy DEMATEL approach	60	3
Kamaruzzaman et al., 2011 [10]	The effect of indoor environmental quality on occupants' perception of performance: A case study of refurbished historical buildings in Malaysia	46	0
Filippi, 2015 [11]	Remarks on the green retrofitting of historic buildings in Italy	43	0
Conejos et al., 2016 [7]	Governance of heritage buildings: Australian regulatory barriers to adaptive reuse	34	3

Note: GS: Global Citations Score, LC: Local Citations Score

the Global Citations Score and the Local Citations Score, respectively. Table 2.2 shows the five most globally cited publications in the database. It is worth noting that two out of these five articles are also among the most locally cited articles [7, 8].

2.4.2   *Collaboration Networks*

Collaboration networks illustrate how the various scientific actors relate to each other in a scientific field of research [6]. In the present study, the collaboration between authors and countries is illustrated via the VOSviewer software package, characterized by being user-friendly and its ability to create clear visualization networks [12].

Coauthorship analysis of authors demonstrates the social knowledge structure of a research field and identifies regular groups of authors and authors with significant contributions in this field [6]. In the 90 extracted publications, a total of 288 authors were identified. Figure 2.3 shows the coauthorship network of the 27 authors having at least two publications and five citations. Each node represents an author, and the node size reflects each author's activeness (number of publications). The coloring scheme illustrated in the legend reflects the average citation per document for the authors. Furthermore, authors are ranked based on Productivity (No. of Publications), Influence (Citation Score), Activeness (Average Publication per Year), and Centrality (Collaboration Links) in Table 2.3.

Several observations can be concluded from the conducted analysis. Only 9% (27/288) of the total authors had at least two publications and five citations. The most productive and central author in the field is Silva A. Their research interests are related to heritage buildings' functionality [13]. Further, Aigwi I. is among the

Figure 2.3 Authors Collaboration Network

Table 2.3 Authors Collaboration Network Indicators

Author	NP	CS	APY	CL
Most productive authors				
Silva, A.	3	40	2017.7	4
Ge, Jian	3	6	2017.7	2
Authors with the highest influence				
Aigwi, Itohan Esther	2	55	2018.5	2
Egbelakin, Temitope	2	55	2018.5	2
Ingham, Jason	2	55	2018.5	2
Chew, Michael Y. L.	2	50	2016.5	1
Conejos, Sheila	2	50	2016.5	1
Most active authors				
Goncalves, Joana	2	6	2021.5	1
Mateus, Ricardo	2	6	2021.5	1
Lu, Jiapan	2	3	2021.5	2
Luo, Xiaoyu	2	3	2021.5	2
Authors with the highest centrality				
Alejandre, F. J.	2	32	2018.0	4
Macias-Bernal, J. M.	2	32	2018.0	4
Prieto, A. J.	2	32	2018.0	4
Silva, A.	3	40	2017.7	4
De Brito, J.	2	22	2017.0	4

Note: NP: Number of Publications, CS: Citations Score, APY: Average Publication Year, CL: Collaboration Links

most influential authors [14], and Goncalves J. is one of the most active authors in SHB [15].

Studying the geographic distribution of a research field is necessary, especially for junior researchers, to locate potential funding opportunities. Out of the total 37 countries in the database, 13 countries had a minimum of three publications and five citations, as shown in Figure 2.4. Like the authors' collaboration network, each node represents a country, and the node size reflects each country's activeness (number of publications). The coloring scheme illustrated in the legend reflects the average citation per document for the countries. The identified 13 countries are ranked in Table 2.4 based on Productivity (No. of Publications), Influence (Citation Score), Activeness (Average Publication per Year), and Scientific Value (Average Citation per document).

The geographic distribution analysis shows that Portugal, Spain, Italy, and China outperform the other countries regarding productivity (60% of total publication count) and citation score. This is natural, given that the European heritage accounts for 30% of the global stock [16]. The most active countries are Netherlands, Poland, and Egypt, with average publication years around end of 2020 and beginning of 2021. It is worth noting that Egypt is witnessing a surge of efforts dedicated to sustainable development in the last few years. In addition, Australia, Italy, and Belgium are superior in their average citations per document (17.7, 14.8,

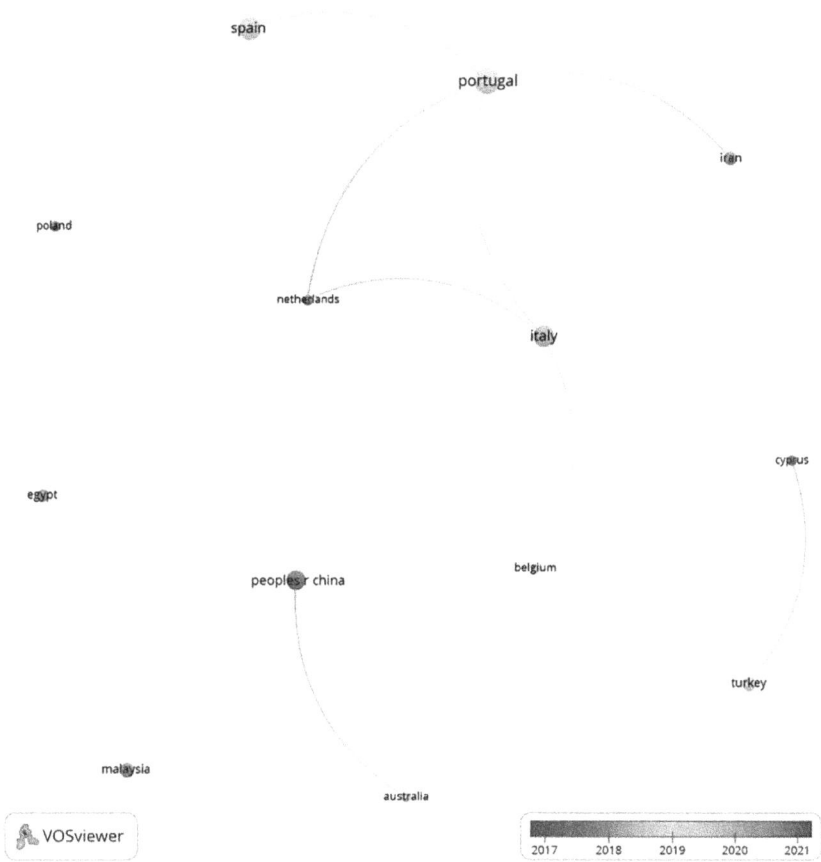

Figure 2.4 Countries Collaboration Network

Table 2.4 Top Contributing Countries

Country	NP	CS	APY	ACD
Portugal	17	154	2019.3	9.1
Spain	14	177	2019.4	12.6
Italy	13	193	2018.2	14.8
Peoples R China	10	100	2017.1	10.0
Malaysia	6	73	2017.4	12.2
Turkey	6	22	2018.6	3.7
Iran	5	18	2017.5	3.6
Egypt	4	22	2020.8	5.5
Australia	3	53	2018.3	17.7
Belgium	3	40	2019.0	13.3
Cyprus	3	40	2013.3	13.3
Netherlands	3	24	2021.3	8.0
Poland	3	5	2021.0	1.7

Note: NP: Number of Publications, CS: Citations Score, APY: Average Publication Year, ACD: Average Citation per Document and 13.3, respectively), demonstrating the significant influence of their academic efforts on the scientific community.

2.4.3   Sources Citation Networks

It is vital to identify the prominent journals of a research domain to distinguish reliable sources of information. Out of the 24 peer-reviewed journals, nine journals had a minimum of three publications and three citations, as shown in the network produced by VOSviewer in Figure 2.5. Similar to the collaboration networks, each node represents a journal, the node size reflects each journal's activeness (number of publications), and the coloring scheme illustrated in the legend reflects the average citation per document for the journals. Table 2.5 lists the top journals ranked based on Productivity (No. of Publications), Centrality (Collaboration Links), Global Citation Score to measure the overall influence of the publications, and Local Citation Score to assess the influence of the publications within the extracted bibliographic database.

The conducted citation analysis revealed some interesting observations about the most productive journals. For example, the top journals "Energy and Buildings, International Journal of Architectural Heritage, Journal of Building Engineering, and Buildings" have 45 publications, which is 50% of the extracted database. Further, only one of those most productive journals has the highest average citations per document (19.3): "Energy and Buildings." Additionally, "Energy and Buildings" and

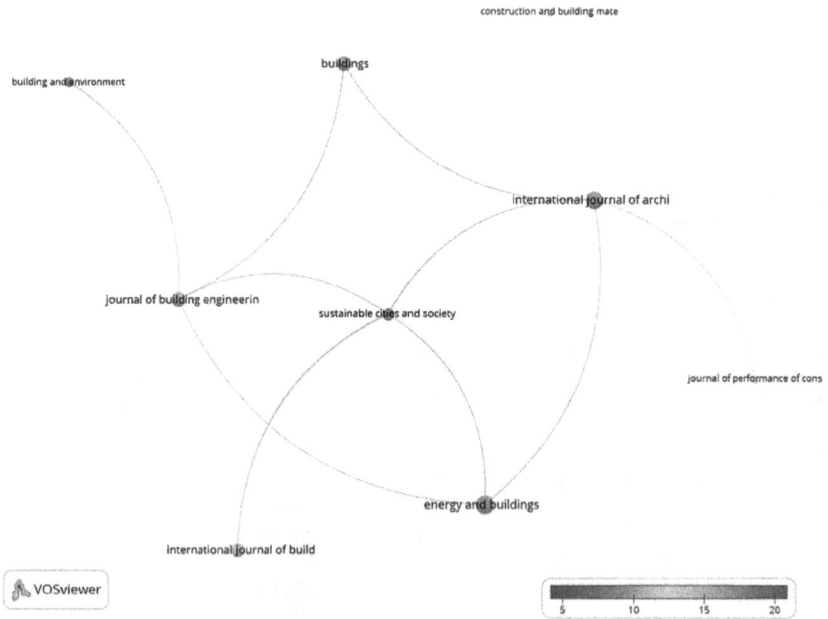

Figure 2.5 Journals Citation Visualization Network

Table 2.5 Top Contributing Journals

Journal	NP	CS	APY	ACD
Energy and Buildings	14	270	2016.3	19.3
International Journal of Architectural Heritage	13	91	2019.0	7.0
Journal of Building Engineering	9	72	2020.3	8.0
Buildings	9	61	2019.9	6.8
International Journal of Building Pathology and Adaptation	8	74	2018.3	9.3
Sustainable Cities and Society	6	118	2019.7	19.7
Building and Environment	4	94	2016.8	23.5
Journal Of Performance of Constructed Facilities	3	36	2016.0	12.0
Construction and Building Materials	3	33	2019.3	11.0

Note: NP: Number of Publications, CS: Citations Score, APY: Average Publication Year, ACD: Average Citation per Document

"International Journal of Architectural Heritage" are superior to other journals, with citation scores of 270 and 91, respectively. Finally, "Journal of Building Engineering," "Buildings," and "International Journal of Architectural Heritage" are the most active in the top contributing journals. Consequently, follow-up research can be conducted by gathering research articles about SHB from these top journals.

2.4.4   Conceptual Networks

Conceptual networks show the relationships between keywords or topics in publications [6]. It helps interested readers and researchers determine the most occurring topics, define the main themes of a research field, and study the trending topics. These goals are accomplished through cooccurrence analysis, cluster analysis, theme analysis, and trend topics analysis over the 20-year study period.

2.4.4.1   Cooccurrence Analysis

Cooccurrence analysis quantitatively constructs the conceptual knowledge structure of a research field through the number of times a term appears in the database articles [17]. A thesaurus file was used in VOSviewer to merge similar terms and avoid duplication (e.g., replace "heritage conservation" and "conservation plan" with "conservation"). Of the total 360 keywords, 15 occurred at least twice in the database. Table 2.6 lists the most occurring keywords in the extracted bibliographic data about SHB.

2.4.4.2   Cluster Analysis

The identified clusters are categorized into five clusters, as shown in Figure 2.6, generated by VOS viewer. Clusters consist of keywords that are found together in

Table 2.6 Top Keywords in Sustainable Heritage Buildings

Keywords	No. of Occurrences
Heritage buildings	40
Sustainability	17
Conservation	13
Adaptive reuse	9
Energy efficiency	7
Maintenance	4
Vernacular architecture	4
Thermal comfort	3
BIM	3
Diagnosis	2
Energy retrofit	2
Environmental maintenance impact (EMI)	2
Life cycle assessment (LCA)	2
Optimization	2
Resilience	2

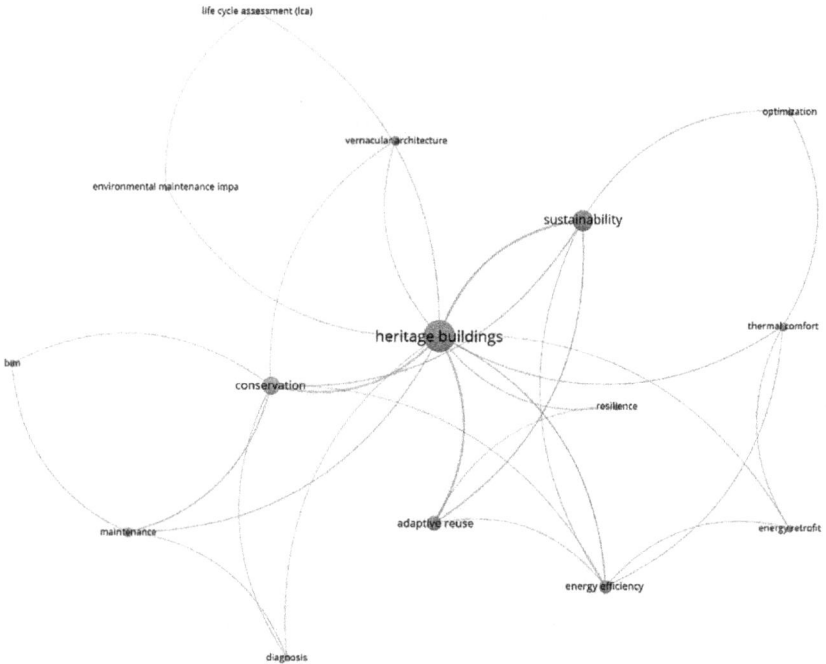

Figure 2.6 Clusters Visualization Network

similar research. Each node stands for a keyword, and its size reflects its number of occurrences. Different clusters are given different node colors, as in Table 2.7. Since only a portion of the total keywords is included in this analysis, the total publications percentage may not necessarily add up to 100%.

Table 2.7 Identified Clusters Information

Cluster ID	Cluster Keywords	Publications
1	Energy efficiency Energy retrofit Optimization Sustainability Thermal comfort	34.4%
2	BIM Conservation Diagnosis Maintenance	24.4%
3	Adaptive reuse Heritage buildings Resilience	12.2%
4	Environmental maintenance impact (EMI) Life cycle assessment (LCA)	4.4%
5	Vernacular architecture	4.4%

2.4.4.3   Theme Analysis

The annual publication trends of the extracted database show that most of the publications were produced between 2015 and 2022. The main themes in these publications were identified using Biblioshiny software introduced by Aria and Cuccurullo [17]. Themes are categorized based on two indices denoted by 2D axes in the strategic thematic map: centrality and density. Centrality reflects the scientific value of themes, while density demonstrates their degree of development [18]. The categories are explained in Table 2.8 [18]. Figure 2.7 illustrates the strategic diagram for the SHB research themes between 2015 and 2022. It can be observed that within the SHB research domain, the motor and basic themes which can be considered the topics with the highest potential for further exploration include "adaptive reuse, conservation, maintenance, diagnosis, thermal comfort, and energy efficiency and retrofit."

As per the analysis of the generated conceptual networks, the most pressing issues to tackle regarding "SHB" are those related to "Conservation, Diagnosis, and Maintenance," "Adaptive Reuse," and "Thermal Comfort, Energy Efficiency, and Energy Retrofit." Alternatively, the results indicate that enhancing heritage buildings' thermal and energy performance requires adaptive reuse and improving their diagnosis, maintenance, and conservation techniques. This is aligned with the United Nations SDGs from the 2030 agenda, especially Goal 11: "Make cities and human settlements inclusive, safe, resilient and sustainable" [1], in addition to the European Green Deal for the recovery of the COVID-19 pandemic [19]. Besides, in response to the global warming and climate change crises, the energy demand for buildings constructed up until 1900 can be decreased by up to 60% with medium retrofits and by 50%–80% with major refurbishments [16].

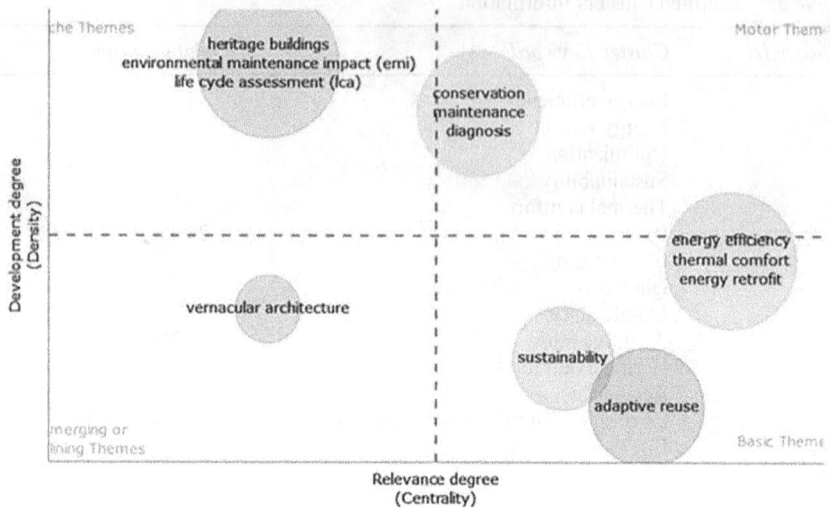

Figure 2.7 Research Themes (2015–2022)

Table 2.8 Description of Research Themes Categories

Category	Location in Diagram	Description
Motor themes	Upper right quadrant	Most developed research areas
Basic/transversal themes	Lower right quadrant	Relevant themes, but not fully developed
Niche themes	Upper left quadrant	Well-developed isolated themes
Emerging/declining themes	Lower left quadrant	Themes starting to develop/decline

2.5   Conclusions

The present study has quantitatively examined the literature conducted about SHB between 2002 and 2022 in 90 peer-reviewed journal publications. The proposed methodology consisted of four steps: (1) define the research objective, (2) select bibliometric approaches, (3) gather bibliometric data, (4) conduct science mapping analysis, and (5) discuss the obtained results. The implemented science mapping analysis benefits several entities interested in the "SHB" research domain. To begin with, junior researchers can identify promising topics, authors to collaborate with, countries with funding opportunities, and best-fit journals for their publications. Moreover, university research groups can direct their efforts toward emerging research topics. Finally, policymakers can find the academic support and expert knowledge required to address the practical implications of SHB.

There are some limitations in the conducted science mapping analysis. For instance, utilizing bibliographic data from one database and the subjective refining criteria may impact the results. Replicating the proposed methodology using more than one database and with different search criteria may produce differing results.

In addition, the scope of this research can be broadened by analyzing the extracted publications' content instead of relying only on their abstracts, keywords, and titles to reach a conclusion.

References

[1] United Nations General Assembly (2015). Transforming our world: The 2030 agenda for sustainable development – A/RES/70/1. https://sustainabledevelopment. un.org/content/documents/21252030 Agenda for Sustainable Development web.pdf. (Accessed November 8, 2022).

[2] Nadkarni, R. R., & Puthuvayi, B. (2020). A comprehensive literature review of multi-criteria decision making methods in heritage buildings. *Journal of Building Engineering*, *32*, 101814.

[3] Posani, M., Veiga, M. D. R., & de Freitas, V. P. (2021). Towards resilience and sustainability for historic buildings: A review of envelope retrofit possibilities and a discussion on hygric compatibility of thermal insulations. *International Journal of Architectural Heritage*, *15*(5), 807–823.

[4] Cabeza, L. F., de Gracia, A., & Pisello, A. L. (2018). Integration of renewable technologies in historical and heritage buildings: A review. *Energy and Buildings*, *177*, 96–111.

[5] Dias Pereira, L., Tavares, V., & Soares, N. (2021). Up-to-date challenges for the conservation, rehabilitation and energy retrofitting of higher education cultural heritage buildings. *Sustainability*, *13*(4), 2061.

[6] Gutiérrez-Salcedo, M., Martínez, M. Á., Moral-Munoz, J. A., Herrera-Viedma, E., & Cobo, M. J. (2018). Some bibliometric procedures for analyzing and evaluating research fields. *Applied Intelligence*, *48*(5), 1275–1287.

[7] Conejos, S., Langston, C., Chan, E. H., & Chew, M. Y. (2016). Governance of heritage buildings: Australian regulatory barriers to adaptive reuse. *Building Research & Information*, *44*(5–6), 507–519.

[8] Vardopoulos, I. (2019). Critical sustainable development factors in the adaptive reuse of urban industrial buildings. A fuzzy DEMATEL approach. *Sustainable Cities and Society*, *50*, 101684.

[9] Rodrıguez Algeciras, J. A., Gómez Consuegra, L., & Matzarakis, A. (2016). Spatial-temporal study on the effects of urban street configurations on human thermal comfort in the world heritage city of Camagüey-Cuba. *Building and Environment*, *101*, 85–101.

[10] Kamaruzzaman, S. N., Egbu, C. O., Zawawi, E. M. A., Ali, A. S., & Che-Ani, A. I. (2011). The effect of indoor environmental quality on occupants' perception of performance: A case study of refurbished historic buildings in Malaysia. *Energy and Buildings*, *43*(2–3), 407–413.

[11] Filippi, M. (2015). Remarks on the green retrofitting of historic buildings in Italy. *Energy and Buildings*, *95*, 15–22.

[12] Van Eck, N. J., & Waltman, L. (2010). Software survey: VOSviewer, a computer program for bibliometric mapping. *Scientometrics*, *84*(2), 523–538.

[13] Prieto, A. J., Macías-Bernal, J. M., Chávez, M. J., Alejandre, F. J., & Silva, A. (2019). Impact of maintenance, rehabilitation, and other interventions on functionality of heritage buildings. *Journal of Performance of Constructed Facilities*, *33*(2), 04019011.

[14] Aigwi, I. E., Egbelakin, T., Ingham, J., Phipps, R., Rotimi, J., & Filippova, O. (2019). A performance-based framework to prioritise underutilised historical buildings for

adaptive reuse interventions in New Zealand. *Sustainable Cities and Society, 48,* 101547.

[15] Gonçalves, J., Mateus, R., Dinis Silvestre, J., Pereira Roders, A., & Vasconcelos, G. (2022). Selection of core indicators for the sustainable conservation of built heritage. *International Journal of Architectural Heritage, 16*(7), 1047–1062.

[16] Tejedor, B., Lucchi, E., Bienvenido-Huertas, D., & Nardi, I. (2022). Non-Destructive Techniques (NDT) for the diagnosis of heritage buildings: Traditional procedures and futures perspectives. *Energy and Buildings, 263,* 112029.

[17] Aria, M., & Cuccurullo, C. (2017). Bibliometrix: An R-tool for comprehensive science mapping analysis. *Journal of Informetrics, 11*(4), 959–975.

[18] Garcia-Buendia, N., Moyano-Fuentes, J., Maqueira-Marín, J. M., & Cobo, M. J. (2021). 22 Years of lean supply chain management: A science mapping-based bibliometric analysis. *International Journal of Production Research, 59*(6), 1901–1921.

[19] European Commission, A European Green Deal. Striving to be the first climate-neutral continent, (2019). https://ec.europa.eu/info/strategy/priorities-2019-2024/european-green-deal_en (accessed November 8, 2022).

3 Maintainability Index of Heritage Buildings

Hoda Abdelrazik

Head of Sector, Ministry of Housing Utilities and Urban Communities, Cairo, Egypt

Mohamed Marzouk

Professor of Construction Engineering and Management, Structural Engineering Department, Faculty of Engineering, Cairo University, Postal code 12613, Giza, Egypt

3.1 General

This chapter presents a procedure that determines the parameters affecting the preservation and maintenance of heritage buildings related to the Egyptian context and climate conditions. The procedure adopts Analytical Hierarchy Process (AHP) as a decision-making technique to determine the weights of identified parameters. A Maintainability Index of Heritage Buildings (MIHB) is introduced using the governing parameters introduced to support the decision-making process of setting priorities for the maintenance of heritage buildings. The maintainability index captures the current status of these buildings and their severity condition. MIHB ranges are introduced from excellent condition to very poor condition. The index is calculated for a heritage case study with an already known condition to demonstrate its use in predicting the condition of a heritage building.

3.2 Introduction

Egypt's heritage is one of the most valuable and diverse in the world, with many monuments belonging to different historical eras alongside the Nile River and adjacent to the Mediterranean and Red Sea shores and in remote desert areas. These buildings link us to the past; they reach back and connect current generations to identity and history, which cannot be easily replaced. Egyptian-built heritage is subject to numerous threats. A lot of buildings are subject to damage. Besides intentional demolition, some are subject to partial collapse due to their declined conditions and severe deterioration. This is a result of several factors, including the influence of time and weather conditions, catastrophic events affecting them, minor or no maintenance procedures, vandalism, encroachment of new building blocks, weak management of heritage sites, lack of funding, and failure to enforce the law.

Many Heritage structures experienced damaging earthquakes and fires across their life spans. Heritage sites are mostly ignored by the local population unless

DOI: 10.1201/9781003357483-3

safeguarded by the government. Saad et al. [1] stated that the restoration of historic buildings located in the Historic Cairo area alone is considered a challenging task since the process involves complex work and huge amounts of funding.

Heritage buildings represent significant financial and cultural wealth for the countries. Governments have come to recognize the significance of the conservation of cultural assets. Recently, an increase in policies and practices concerned with preserving and conserving heritage buildings has been witnessed. The concept of conservation of cultural built heritage has developed internationally over recent years to define comprehensive measures for interventions that will guide to the optimum preservation required [2].

In the initial process of conservation of heritage buildings, the building should be studied from all aspects to protect its structural system and aesthetic characteristics, maintain its function, acknowledge current lifestyles, and prevent its degradation [3, 4]. The conservation of heritage buildings is a complex task and is usually based on a more comprehensive analysis of various criteria not limited to artistic or cultural. As referred to by UNESCO [5], it requires evaluating a number of factors, which requires knowledge of buildings' history and present to foresee their future. The difficulty of conservation management tasks is attributed to technical, financial, and legislative aspects [6]. Kim et al. [7] mentioned that the absence of accurate decision tools and decision support systems that can aid in determining conservation priorities leads to weak tools for prioritizing tasks and consequently applying inefficient conservation and maintenance procedures.

Haagenrud et al. [8] stated that different factors cause the deterioration of buildings. These factors cause direct and indirect requirements regarding building maintenance and repair costs. He emphasized that the service life of heritage buildings is an important parameter that contributes to countries' social and economic strength. Over time, the gradual decay of architectural heritage influences society's needs and expectations. Forsyth [9] explained that a heritage building is believed to be valuable because of its age, unique, and variation from the norm. This establishes the heritage building's importance and genuineness, an important quality to retain. He also explained that the structure is the basis of the building's shape and is what displays the building's aesthetic significance. He stated that to understand the condition of a structure, accurate numerical and qualitative analysis have to be performed together with the study of its historical records to understand the development of its behavior with time. The current methodologies in structural design differ from the past in that experience and performance have been standardized, parameterized, rationalized, and provided with a scientific basis. Thus, current codes of practice and standards apply to structures that predate these codes. Warren [10] also stated that the main determinant of a building's shape is its structure, thus establishing its aesthetic value. He indicated that it is important to understand the condition of the structure in order to preserve its elegance and style. Friedman [11] pointed out that heritage building conservation is the use of techniques and analysis in the appraisal of the structural fitness of elements of the structure, taking into account that this structure was originally not constructed according to our engineering codes.

Today, the Venice Charter [12] regulates the preservation of historic buildings worldwide. The Charter differentiates between conservation and restoration by expressing that the principle of conservation is to maintain, while restoration addresses the historic and aesthetic value. The ICOMOS Bura charter [13] explained that maintenance is vital to conservation. The Charter explained that conservation should be accepted when the fabric is of cultural significance, and its maintenance is necessary to retain that significance. It recommended conducting appropriate studies (physical, documents, oral and other evidence, drawing from knowledge, skills, and disciplines) before embarking on restoration works.

In Egypt, laws and legislations that govern and protect architectural heritage were also introduced. The antiquities protection law (No. 117 of 1983) [14] protects ancient monuments. Law No. 119 of 2008 [15] regulates the works of the National Organization of Urban Harmony and Areas of Outstanding Value. Law No. 144 of 2006 [16] determines heritage buildings and regulates the demolition of unthreatened buildings and the conservation of architectural heritage. Buildings defined by this law have distinct architectural styles (architectural value), have a connection to the country's history (historical value), are related to historical personalities (symbolic value), represent an exact historical period (historical value), or are used as touristic destinations (social and functional value). This law protects these types of buildings.

Alba-Rodríguez et al. [17] studied key criteria for decision-making regarding buildings' renovation (investment costs, building performance conditions, and existing regulations). They also pointed out that theory and practice are usually explored separately although they are interconnected in the case of heritage buildings. The study concluded that conservation budgets should not be the only parameter considered in urban rehabilitation strategies. Rahman et al. [18] identified 24 problems that affect the maintenance of heritage buildings classified into six major classes: (1) financial, (2) spare parts, (3) technical problems, (4) human behavior and attitudes, (5) management and administration, and (6) education and training. The study concluded that the identified problems affect the effectiveness of heritage building maintenance in Malaysia. The research used the importance index method and identified financial problems as the factor influencing other problems.

Sodangi et al. [19] studied 12 environmental factors threatening the service life of heritage buildings in Nigeria. The study identified flood, fire, urban development, and biological agents as the most influencing factors affecting the survival of heritage buildings. Sodangi et al. [20] studied the effect of 26 criteria affecting the maintenance management practices for heritage buildings in Malaysia to produce a set of criteria observed as the most important by curators, academics, and consultants in the maintenance management of heritage buildings. Using the relative importance index method, the study identified the three most important criteria: maintenance staff training and expertise, financial planning and budgets, and conservation plan.

Ortiz et al. [21] applied vulnerability and service life indices using artificial intelligence and the Delphi method to prioritize required preventive measures needed for heritage buildings using weighted variables (materials, structure, anthropogenic factors, and ventilation) as well as hazards (static, structural, environmental, and anthropogenic). Prieto et al. [22] evaluated the interventions performed on 390 historic structures and their relative effect on these buildings' functionality. The study used a developed fuzzy model to evaluate the efficiency of these interventions on the performance of these structures. Thus, identifying the most frequently used maintenance procedures and their impact on the buildings and predicting the future behavior of the analyzed buildings. Vecco et al. [23] developed a cultural heritage sustainability index for conflict-affected regions, focusing on active war areas. The index was developed using information from UNESCO for 207 countries worldwide. The index was calculated for 2008 only because of the limitations of existing cultural statistics. The index displayed the adverse changes likely to affect socioeconomic conditions of cultural heritage in the studied areas and identified regions likely to be further affected by war. The study found that the value of the index is negatively and strongly related to tourism trips and other cultural events.

Whitman et al. [24] studied the thermal improvement of historic timber-framed buildings in the United Kingdom. They found that when considering the energy retrofit of heritage buildings, it is essential to recognize both the technical issues involved and the probable effect on their cultural value. Ruiz-Jaramillo et al. [25] developed a heritage risk index to prioritize public historic preservation tasks. A set of indicators is identified and used to evaluate different components of a heritage building. The index defines the risks related to different aspects, including the building's safety, stability, and habitability, thus enabling the decision-maker to compare different buildings. Using such a method enables managers to create a hierarchical map of risk and decide the priorities of intervention, producing sustainable and critical management restoration and maintenance plans. Haroun et al. [26] developed a Multi-Criteria Decision-Making tool to tackle complex decision-making difficulties which arise when choosing the reuse alternatives for heritage buildings. The criteria and weights used in the evaluation can be modified for different buildings to reflect the variations in purposes and environmental limitations.

Saad et al. [1] presented a framework of a newly developed multiobjective fund-allocation optimization model for prioritizing maintenance actions for historic buildings considering funds availability, structural physical performance, and socioeconomic benefits. Marzouk et al. [27] developed an expert system to select the most effective technique for repairing heritage building materials and recommend suitable procedures to repair different types of deterioration, particularly three identified major problematic elements: timber doors, iron gates, and ceiling paintings .

3.3   Parameters Identification

Parameters affecting the decision-making process of maintenance of heritage buildings in Egypt were collected from studies performed prior to maintenance of some heritage buildings in Egypt, standards, codes of practice, books, academic papers, and interviews with the stakeholders and consultants with a long experience in the field of conservation in Egypt. Sixty-three parameters were identified and divided into six groups: 5 cultural parameters, 15 architectural parameters, 8 geotechnical parameters, 19 structural parameters, 7 material parameters, and 10 external parameters. A questionnaire was distributed among experts to obtain their opinions concerning the importance of these parameters. Detailed description of the identified 63 parameters can be found elsewhere [28, 29].

3.4   Parameters Ranking

Relative Importance Index (RII) is a widely used technique for analyzing structured questionnaire responses for data involving ordinal measurement of attitudes. It is used to describe the relative importance of specific causes and effects based on their likelihood of occurrence and effect using the Likert scale of five scales. The higher values of the index of relative importance (RII) are the significant causes or influential components. The relative importance index (RII) was calculated using Equation 3.1 (Sambasivan and Soon [30]).

$$RII = \frac{\sum W}{A * N} \tag{3.1}$$

where W is the weight given to each parameter by the respondents (ranging from 1 to 5), A is the highest weight (i.e., 5 in this case), and N is the total number of respondents.

According to the highest RII values, the top relative importance values range from 0.82 to 0.97 and capture 21 parameters, as listed in Table 3.1. The results concluded that the most significant group of parameters is the Geotechnical group with a relative importance index of 0.9, followed by the structural group with an index of 0.82 and then the cultural group with an index of 0.8.

3.5   Identifying Parameters Weights

The AHP developed by Thomas Saaty in the 1970s is considered one of the most common tools of multicriteria decision-making methods. One of its advantages is its ease of use due to its hierarchical structure, which can be easily altered in size to accommodate decision-making problems. The highest-ranking parameters from the relative importance index analysis were grouped according to their categories and divided into four hierarchy levels, as shown in Figure 3.1.

A questionnaire was prepared according to the previously mentioned levels and distributed among experts in the field of conservation and maintenance of heritage building projects which represent owners (government bodies mainly concerned with the maintenance of historical palaces) and consultants involved in the design and supervision of conservation and maintenance works including the experts interviewed in the first questionnaire. The input table for pairwise comparisons was sent to participants for their input judgments, where ten matrices were provided: one matrix for the main category and nine matrices for all sub-criteria.

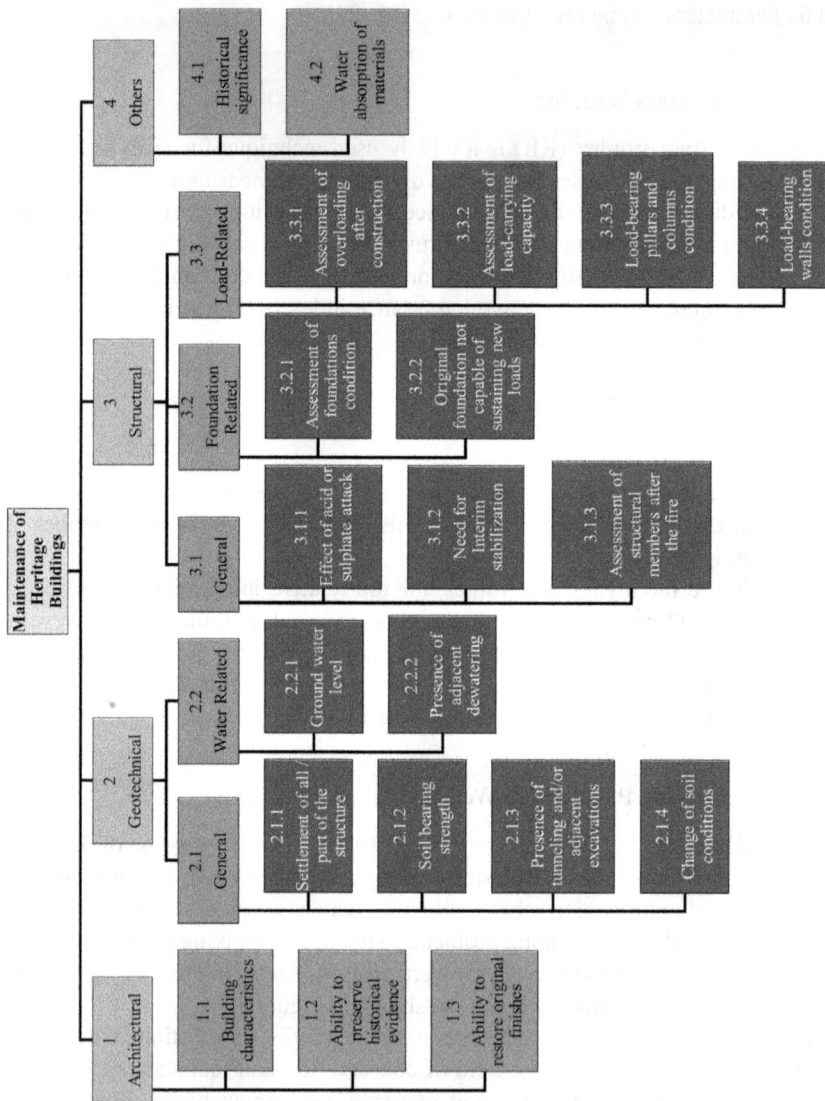

Figure 3.1 Hierarchy for Governing Maintenance Parameters

The participants were asked to compare the relative importance of the parameters by specifying the intensity based on the scale of 1–9 suggested by Saaty [31].

The feedback received from 14 experts is fed to the AHP Excel Template with Multiple Inputs developed by Klaus D. Goepel – Business Performance Management Singapore – licensed under the Creative Commons Attribution-Noncommercial 3.0 Singapore License and its revisions [32, 33]. The spreadsheet template does not include the hierarchy of the decision problem and the final aggregation of weights. Thus, it is only suitable for finding the weights in each category or sub-category. Once participants provide their judgments and pairwise matrices are produced, individual judgments are aggregated into group judgments. Consistency checking ensures logical judgment with an acceptable value of less than 0.1. One feedback response from an expert was excluded due to inconsistency. In order to evaluate the degree of agreement between different participants, an acceptable level of consensus ensures that the final results are based on the collective opinions of the group of experts. Values below 50% indicate no consensus within the group and a high diversity of judgments. Values in the 80%–90% range indicate a high overlap of priorities and excellent agreement of judgments from the group members.

The feedback received from experts for the four main parameter groups (architectural, geotechnical, structural, and others) was fed into the AHP spreadsheet. The feedback from 13 experts produced the weights shown in Figure 3.2 with a consistency ratio of 0.013. The consensus ratio showed a percentage of 81%, indicating excellent agreement of judgments from respondents, as shown in the figure. The matrix also shows the dominance of the Structural parameters group with a percentage of 44.2%, followed by the Geotechnical parameters group with a percentage of 38.4 %, the Architectural group was given a percentage of 11.6%, and the Others group scored a percentage of 5.9%. Similarly, the weights are estimated for the rest of the parameters that exist on the same level. The weights of the four levels of AHP matrices are aggregated and listed in Table 3.2, which shows the final weights of parameters based on expert opinions and using the Analytical Hierarchy Process.

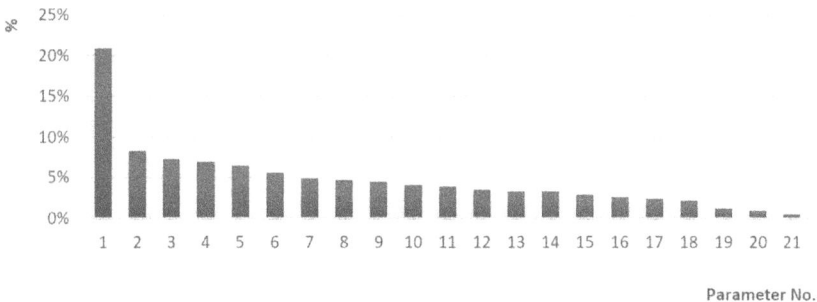

Figure 3.2 Percentage of Contribution of Parameters in MIHB

Table 3.1 Estimated Top RII Values for Parameters

No.	Group	Parameter	RII
1	Geotechnical	Settlement of all/part of the structure	0.97
2	Architectural	Building characteristics	0.94
3	Geotechnical	Soil bearing strength	0.92
4	Geotechnical	Ground water level	0.90
5	Structural	Assessment of the foundations condition	0.90
6	Cultural	Historical significance	0.89
7	Geotechnical	Presence of adjacent dewatering	0.89
8	Architectural	Ability to preserve historical evidence	0.88
9	Geotechnical	Presence of tunneling and/or adjacent excavations	0.88
10	Geotechnical	Change of soil conditions	0.88
11	Structural	Assessment of overloading after original construction	0.88
12	Structural	Original foundation not capable of sustaining new loads in previous extensions	0.88
13	Structural	Assessment of load-carrying capacity	0.86
14	Structural	Load-bearing pillars and columns condition	0.86
15	Structural	Effect of acid or sulfate attack	0.85
16	Structural	Need for Interim stabilization during restoration	0.85
17	Structural	Load-bearing walls condition	0.84
18	Structural	Assessment of structural members after the fire	0.84
19	Materials	Water absorption of materials causing erosion	0.84
20	Architectural	Ability to restore original finishes	0.82
21	Geotechnical	Change of water regime	0.82

3.6   Maintainability Index of Heritage Buildings

MIHB is introduced to support the decision-making process of setting priorities for the maintenance of heritage buildings. The maintainability index captures the current status of these buildings and their severity condition. MIHB ranges are introduced from excellent condition to very poor condition. A grading scheme is proposed through a set of criteria that measure varying levels of building performance conditions. The scheme was divided into five ranges of conditions to make it easier for the decision maker to predict a nearly precise estimate of the condition of the heritage building, as listed in Table 3.3. The condition ranges are as follows:

- Excellent condition ranges from 81% to 100%, where vulnerabilities and hazards are very low, and maintenance action is not required, which might be the case for a recently maintained building.
- Very good condition ranges from 66% to 80%, where it is assumed that vulnerabilities and hazards are low, and maintenance is not necessary.
- Good condition, which ranges from 46% to 65% and assumes that vulnerabilities and hazards are moderate, where costs, benefits, and safety are considered and balanced to decide when it is necessary to intervene.

- Poor condition ranges from 21 to 45 percent and assumes that vulnerabilities and hazards are high and possible intervention in a short period is recommended.
- Very poor condition ranges from 1% to 20%, and assumes that vulnerabilities and hazards are especially aggressive and intervention in a short period is essential.

Using the heritage buildings available data, recorded maintenance actions, and visual inspection and testing, it is possible to determine the most frequent anomalies acting on the building and their causes. The vulnerable elements responsible for the problems can be pointed out. This allows for the understanding of the weak points of these buildings. The analysis of maintenance and conservation works data may enable the decision-maker to identify the optimum time for implementing maintenance actions, thus reducing maintenance costs and promoting the sustainability of conservation policies.

Table 3.2 Final Parameter Weights

Level 1	Level 2	Level 3		Matrix Weight	Final Weight
1	Architectural			0.116	
	1.1	Building characteristics		0.381	0.044
	1.2	Ability to preserve historical evidence of the building		0.338	0.039
	1.3	Ability to restore original finishes		0.281	0.033
2	Geotechnical			0.384	
	2.1	General		0.536	
		2.1.1	Settlement of all/part of the structure	0.2	0.041
		2.1.2	Soil bearing strength	0.24	0.049
		2.1.3	Presence of tunneling and/ or adjacent excavations	0.227	0.047
		2.1.4	Change of soil conditions	0.333	0.069
	2.2	Water related		0.464	
		2.2.1	Ground water level	0.371	0.066
		2.2.2	Presence of adjacent dewatering	0.464	0.083
		2.2.3	Change of water regime	0.165	0.029

(*Continued*)

Table 3.2 (Continued)

Level 1	Level 2	Level 3		Matrix Weight	Final Weight
3	Structural			0.442	
	3.1	General		0.085	
		3.1.1	Effect of acid or sulfate attack	0.255	0.009
		3.1.2	Need for Interim stabilization during restoration	0.151	0.006
		3.1.3	Assessment of structural members after the fire	0.594	0.022
	3.2	Foundation related		0.632	
		3.2.1	Assessment of the foundations condition	0.748	0.209
		3.2.2	Original foundation not capable of sustaining new loads in previous extensions	0.252	0.070
	3.3	Load related		0.283	
		3.3.1	Assessment of overloading after construction	0.192	0.024
		3.3.2	Assessment of load-carrying capacity	0.272	0.034
		3.3.3	Load-bearing pillars and columns condition	0.441	0.055
		3.3.4	Load-bearing walls condition	0.095	0.012
4	Others			0.058	
	4.1	Historical significance		0.567	0.033
	4.2	Water absorption of materials causing erosion		0.433	0.025
Total					1.000

MIHB is proposed. The index utilizes the governing parameters to help set priorities for maintenance actions. The condition ranges proposed in Table 3.3 result in five values for each parameter to predict the heritage building condition, as shown in Table 3.4. The values for MIHB are considered the mid-point for the different conditions. MIHB is considered 0.9, 0.7, 0.55, 0.3, and 0.1

Table 3.3 Severity Conditions

Condition	MIHB	Condition Explanation
Excellent condition	[81–100]	The vulnerabilities and hazards are regarded as very low, and an intervention or maintenance action is not required for a long period.
Very Good condition	[66–80]	The vulnerabilities and hazards are regarded as low, and maintenance is not required for a short-medium period.
Good condition	[46–65]	The vulnerabilities and hazards are moderate, where costs and benefits are considered and balanced to decide when it is necessary to intervene
Poor condition	[21–45]	The vulnerabilities and hazards are regarded as high, and a probable intervention in a short period is recommended
Very Poor condition	[01–20]	The vulnerabilities and hazards are regarded as especially aggressive, and intervention in a short period is essential

for Excellent, Very Good, Good, Poor, and Very Good, respectively. Table 3.5 lists the weighted condition values for the 21 parameters considered. The overall MIHB aids in making comparisons among different heritage buildings. The index is a dimensionless coefficient that indicates the level or severity of the damage to the heritage asset.

Figure 3.2 indicates that the first ten parameters contribute by 74% in MIHB. In comparison, the first 15 and 18 parameters contribute by 88% and 98% in MIHB. The last three parameters (load-bearing walls condition, effect of acid or sulfate attack, and need for interim stabilization during restoration) have a 2% contribution in MIHB. This means that neglecting such parameters would not affect the results.

3.7   Case Study

Makram Ebeid Villa is considered a case study, a two-story wall-bearing building built in 1942 for the notable politician and minister in his hometown in Qena Governorate. The building was subject to intentional sabotage; it survived two fires in its life span, there is a collapse in the entrance hall and part of the stairs, and some parts of the roof balconies also collapsed, as shown in Figure 3.3. There is a collapse in the west and south load-bearing walls due to excavations around the building structure, and there is also a source of damp affecting the foundations, which lead to the settlement of parts of the building. The condition of the building is evaluated to calculate the MIHB as listed in Table 3.6. The results indicate that MIHB has a value of 0.245, a Poor Condition.

Table 3.4 Maintainability Index Weights According to Different Conditions

No.	Parameter	Excellent	Very Good	Good	Poor	Very Poor
		0.9	0.7	0.55	0.3	0.1
1	Assessment of the foundations condition	Excellent	Very good	Good	Need minor stabilization	Need major stabilization
2	Presence of adjacent dewatering	None	None	Small effect	Large effect	Very large effect
3	Ground water level	No change	No change	Small effect	Large effect	Very large effect
4	Original foundation not capable of sustaining new loads in previous extensions	Not applicable	Not applicable	Minimal effect	Need minor stabilization	Need major stabilization
5	Change of soil conditions	No change	No change	Small impact	High impact	Very high impact
6	Load-bearing pillars and columns condition	Excellent condition	Very good condition	Good condition	Need minor repairs	Need major repairs
7	Soil bearing strength	Very high	High	Medium	Low	Very low
8	Presence of tunneling and/or adjacent excavations	No change	No change	Minor impact	Large impact	Large impact
9	Building characteristics	Exceptional value	High value	Good	Medium	Small
10	Settlement of all/part of the structure	No change	No change	Minimal impact	Large impact	Very large impact
11	Ability to preserve historical evidence	Very easy	Easy	Average	Hard to perform	Very hard to perform
12	Assessment of load-carrying capacity	Excellent condition	Very good condition	Good condition	Need minor stabilization	Need major stabilization
13	Ability to restore original finishes	Very easy	Easy	Medium	Hard to perform	Very hard to perform
14	Historical significance	Exceptional value	High value	Good value	Medium value	Small value
15	Change of water regime	No change	Minimal change	Medium impact	Large impact	Major impact
16	Water absorption of materials causing erosion	None	Minimal	Medium impact	Large impact	Very large impact
17	Assessment of overloading after construction	None	Very small	Medium effect	Large effect	Very large impact
18	Assessment of structural members after the fire	Excellent condition	Very good condition	Need minor repairs	Need minor repairs	Need major repairs
19	Load-bearing walls condition	Excellent condition	Very good condition	Good condition	Need minor repairs	Need major repairs
20	Effect of acid or sulfate attack	No effect	Minimal effect	Medium impact	Large impact	Very large impact
21	Need for Interim stabilization during restoration	No need	No need	Minor stabilization	Major stabilization	Major stabilization

Table 3.5 Maintainability Index Values

No.	Parameter	AHP Weights	Excellent	Very Good	Good	Poor	Very Poor
			0.9	0.7	0.55	0.3	0.1
1	Assessment of the foundations condition	0.209	0.188	0.146	0.115	0.063	0.021
2	Presence of adjacent dewatering	0.083	0.075	0.058	0.046	0.025	0.008
3	Ground water level	0.072	0.065	0.05	0.040	0.022	0.007
4	Original foundation not capable of sustaining new loads in previous extensions	0.070	0.063	0.049	0.039	0.021	0.007
5	Change of soil conditions	0.064	0.058	0.045	0.035	0.019	0.006
6	Load-bearing pillars and columns condition	0.055	0.05	0.039	0.030	0.017	0.006
7	Soil bearing strength	0.049	0.044	0.034	0.027	0.015	0.005
8	Presence of tunneling and/or adjacent excavations	0.047	0.042	0.033	0.026	0.014	0.005
9	Building characteristics	0.044	0.04	0.031	0.024	0.013	0.004
10	Settlement of all/part of the structure	0.041	0.037	0.029	0.023	0.012	0.004
11	Ability to preserve historical evidence	0.039	0.035	0.027	0.021	0.012	0.004
12	Assessment of load-carrying capacity	0.034	0.031	0.024	0.019	0.01	0.003
13	Ability to restore original finishes	0.033	0.03	0.023	0.018	0.01	0.003
14	Historical significance	0.033	0.03	0.023	0.018	0.01	0.003
15	Change of water regime	0.029	0.026	0.02	0.016	0.009	0.003
16	Water absorption of materials causing erosion	0.025	0.023	0.018	0.014	0.008	0.003
17	Assessment of overloading after construction	0.024	0.022	0.017	0.013	0.007	0.002
18	Assessment of structural members after the fire	0.022	0.02	0.015	0.012	0.007	0.002
19	Load-bearing walls condition	0.012	0.011	0.008	0.007	0.004	0.001
20	Effect of acid or sulfate attack	0.009	0.008	0.006	0.005	0.003	0.0009
21	Need for Interim stabilization during restoration	0.006	0.005	0.004	0.003	0.002	0.0006

Figure 3.3 Collapse in Makram Ebeid Villa Elements

Table 3.6 MIHB Values for Makram Ebeid Villa

No.	Parameter	Condition	Index
1	Assessment of the foundations condition	Very poor	0.021
2	Presence of adjacent dewatering	Large effect	0.008
3	Ground water level	Very large effect	0.007
4	Original foundation not capable of sustaining new loads in previous extensions	Minimal effect	0.039
5	Change of soil conditions	Very high impact	0.006
6	Load-bearing pillars and columns condition	Need major repairs	0.006
7	Soil bearing strength	Low	0.015
8	Presence of tunneling and/or adjacent excavations	Large impact	0.005
9	Building characteristics	Good	0.024
10	Settlement of all/part of the structure	Very large impact	0.004
11	Ability to preserve historical evidence	Easy	0.027
12	Assessment of load-carrying capacity	Need minor stabilization	0.010
13	Ability to restore original finishes	Easy	0.023
14	Historical significance	Good value	0.018
15	Change of water regime	Major impact	0.003
16	Water absorption of materials causing erosion	Large impact	0.008
17	Assessment of overloading after construction	Large effect	0.007
18	Assessment of structural members after the fire	Poor	0.007
19	Load-bearing walls condition	Needs minor repairs	0.004
20	Effect of acid or sulfate attack	Large impact	0.003
21	Need for Interim stabilization during restoration	Needs major stabilization	0.0006
	Maintainability index of heritage buildings (MIHB)		0.2456

3.8   Conclusion

This chapter presented a framework to support stakeholders in setting priorities for maintaining heritage buildings in Egypt. Significant parameters affecting the current situation of heritage buildings are investigated. The AHP technique was used to determine the relative weights of the 21 top ranked parameters identified from the relative importance index method using pairwise comparisons. MIHB was introduced to capture these buildings' current status and severity conditions. The proposed MIHB takes into consideration dominating Geotechnical, Structural, Cultural, and Architectural aspects of the historic building. A case study was presented to demonstrate the main features of the proposed framework and illustrate its capabilities.

The index describes the state of conservation of a heritage building and the priority for its intervention; it also allows comparison between different buildings. This approach can establish an order of priority for the conservation interventions in the built heritage of a certain municipality. It can also enable decision-makers to understand the weak points of these buildings, which should be carefully analyzed during periodic inspections. The analysis of these data may demonstrate the success that can be achieved by identifying the best time to undertake preventive maintenance actions, thus reducing maintenance costs and promoting the sustainability of conservation policies.

References

[1] Saad, D. A., Elyamani, A., Hassan, M., & Mourad, S. (2019). A fund-allocation optimization framework for prioritizing historic structures' conservation projects-an application to historic Cairo. CSCE Annual Conference, June 12–15, 2019 Laval (Greater Montreal), Quebec, Canada.

[2] International Council on Monuments Sites (ICOMOS). Principles for the analysis, conservation and structural restoration of architectural heritage. Guidelines, ICOMOS, Victoria Falls, Zimbabwe, 2003.

[3] ICOMOS. European charter of the architectural heritage. Adopted by the Council of Europe, 1975.

[4] Ornelas, C., Guedes, J. M., & Breda-Vázquez, I. (2016). Cultural built heritage and intervention criteria: A systematic analysis of building codes and legislation of Southern European countries. *Journal of Cultural Heritage*, 20, 725–732.

[5] UNESCO. Convention concerning the protection of the World cultural and natural heritage, United Nations Educational, Scientific and Cultural Organization, Paris, 1972.

[6] Mickaityte, A., Zavadskas, E. K., Kaklauskas, A., & Tupenaite, L. (2008). The concept model of sustainable buildings refurbishment. *International Journal of Strategic Property Management*, 12(1), 53–68.

[7] Kim, C. J., Yoo, W. S., Lee, U. K., Song, K. J., Kang, K. I., & Cho, H. (2010). An experience curve-based decision support model for prioritizing restoration needs of cultural heritage. *Journal of Cultural Heritage*, 11(4), 430–437.

[8] Haagenrud, S. E. (2004). Factors causing degradation. Guide and bibliography to service life and durability research for buildings and components, Joint CIB W80/RILEM TC 140. *Prediction of Service Life of Building Materials and Components,* 2(1–2), 105.

[9] Forsyth, M. (Ed.). (2007). *Structures & construction in historic building conservation.* London: Blackwell, (3).

[10] Warren, J. (2004). Conservation of structure in historic buildings. *Journal of Architectural Conservation,* 10(2), 39–49.

[11] Friedman, D. (2001). Methodology of conservation engineering. *Journal of Architectural Conservation,* 7(2), 49–63.

[12] Charter, V. (1964, May). International charter for the conservation and restoration of monuments and sites. In IInd International Congress of Architects and Technicians of Historic Monuments, Venice, 25–31.

[13] ICOMOS. (1999). The Burra Charter: The Australia ICOMOS charter for places of cultural significance. Burwood: Australia International Council on Monuments and Sites.

[14] Law no. 117 of 1983 as amended by Law no. 3 of 2010 promulgating the antiquities' protection law published in the official Gazette (6A) in14/2/2010.

[15] Law No.119 of 2008. The Building Law: Alwakae Amasryia, (19A), 2–59. In 1115/2008.

[16] Law 144 (2006). The Executive Law 144 for the demolition of buildings and heritage conservation. Egypt: Alwakae Amasryia, (28a) (248), 5-17.15/7/2006.

[17] Alba-Rodríguez, M. D., Machete, R., Gomes, M. G., Falcão, A. P., & Marrero, M. (2021). Holistic model for the assessment of restoration projects of heritage housing. Case studies in Lisbon. *Sustainable Cities and Society,* 67, 102742.

[18] Rahman, M. A. A., Akasah, Z. A., Abdullah, M. S., Musa, M. K., & Sekitar, U. T. H. O. M. (2012). Issues and problems affecting the implementation and effectiveness of heritage buildings maintenance. *The International Conference on Civil and Environmental Engineering Sustainability (IConCEES 2012),* Johor Bahru, Malaysia.

[19] Sodangi, M., Idrus, A., Khamidi, M. F., & Adam, D. E. (2011). Environmental factors threatening the survival of heritage buildings in Nigeria. *South Asia Journal of Tourism and Heritage,* 4(2), 38–53.

[20] Sodangi, M., Khamdi, M. F., Idrus, A., Hammad, D. B., & Ahmed Umar, A. (2014). Best practice criteria for sustainable maintenance management of heritage buildings in Malaysia. *Procedia Engineering,* 77, 11–19.

[21] Ortiz, R., Macias-Bernal, J. M., & Ortiz, P. (2018). Vulnerability and buildings service life applied to preventive conservation in cultural heritage. *International Journal of Disaster Resilience in the Built Environment,* 9(1), 31–47

[22] Prieto, A. J., Macías-Bernal, J. M., Chávez, M. J., Alejandre, F. J., & Silva, A. (2019). Impact of maintenance, rehabilitation, and other interventions on functionality of heritage buildings. *Journal of Performance of Constructed Facilities,* 33(2), 04019011.

[23] Vecco, M., & Srakar, A. (2018). The unbearable sustainability of cultural heritage: An attempt to create an index of cultural heritage sustainability in conflict and war regions. *Journal of Cultural Heritage,* 33, 293–302.

[24] Whitman, C. J., Prizeman, O., Walker, P., & Gwilliam, J. A. (2019). Heritage retrofit and cultural empathy; A discussion of challenges regarding the energy performance of historic UK timber-framed dwellings. *International Journal of Building Pathology and Adaptation,* 38 (2), 386–404.

[25] Ruiz-Jaramillo, J., Muñoz-González, C., Joyanes-Díaz, M. D., Jiménez-Morales, E., López-Osorio, J. M., Barrios-Pérez, R., & Rosa-Jiménez, C. (2020). Heritage risk index: A multi-criteria decision-making tool to prioritize municipal historic preservation projects. *Frontiers of Architectural Research,* 9(2), 403–418.

[26] Haroun, H. A. A. F., Bakr, A. F., & Hasan, A. E. S. (2019). Multi-criteria decision making for adaptive reuse of heritage buildings: Aziza Fahmy Palace, Alexandria, Egypt. *Alexandria Engineering Journal*, 58(2), 467–478.

[27] Marzouk, M., ElSharkawy, M., Elsayed, P., & Eissa, A. (2020). Resolving deterioration of heritage building elements using an expert system. *International Journal of Building Pathology and Adaptation*, 38(5), 721–735.

[28] Abdelrazik, H., & Marzouk, M. (2021). Investigating parameters affecting maintenance of heritage buildings in Egypt. *International Journal of Building Pathology and Adaptation*, 39(5), 734–755. DOI 10.1108/IJBPA-09-2020-0078.

[29] Abdel Raouf, H. (2022) "Framework for Maintaining Heritage Buildings in Egypt" Ph.D. Thesis, Cairo University, Cairo, Egypt.

[30] Sambasivan, M., and Soon, Y. W. (2007). Causes and effects of delays in Malaysian construction industry. *International Journal of Project Management*, 25(5), 517–526.

[31] Saaty, T. L. (2003). Decision-making with the AHP: Why is the principal eigenvector necessary. *European Journal of Operational Research*, 145(1), 85–91.

[32] Goepel, K. D. (2013, June). Implementing the analytic hierarchy process as a standard method for multi-criteria decision making in corporate enterprises–a new AHP excel template with multiple inputs. In *Proceedings of the International Symposium on the Analytic Hierarchy Process*, 2(10), 1–10). Creative Decisions Foundation Kuala Lumpur.

[33] Goepel, K. D. (2018). Implementation of an online software tool for the analytic hierarchy process (AHP-OS). *International Journal of the Analytic Hierarchy Process*, 10(3), 469–487. https://doi.org/10.13033/ijahp.v10i3.590.

4   Heritage Conservation and Reusability

Maryam ElSharkawy

Instructor, Faculty of Engineering, Cairo University, Postal code
12613, Giza, Egypt

Mohamed Marzouk

Professor of Construction Engineering and Management, Structural
Engineering Department, Faculty of Engineering, Cairo University,
Postal code 12613, Giza, Egypt

4.1   General

The chapter reviews the parameters that influence the need for heritage conservation and reusability. Current initiatives to recognize the importance of conservation to countries and their cultural heritage are being valued worldwide. The continuous development of codes and standards that regulate the preservation of heritage buildings has always been perceived as valuable progress to the Heritage conservation fundamentals proposed by ICOMOS and UNESCO, which framed the reusability of heritage assets. The reusability of heritage buildings is critical to creating and surviving sustainable communities. However, major influential factors influence the decision of the reusability purpose related to environmental, socioeconomic, and political goals. The adaptability of a heritage asset should be assessed according to the involved added value and significant opportunities. Thus, different methods are investigated to evaluate the viability of heritage conservation and reusability.

4.2   Introduction

The current state of the art in Heritage Building Information Modeling (HBIM) has emphasized the significance of conservation, repair, and maintenance of previously built carbon-foot prints rather than dealing with newly built ones with their BIM-associated risks. This chapter covers a literature review concerned with assessing daylight performance using HBIM, introducing HBIM-related technologies that include laser scanning, point cloud data, and virtual reality dealing with different existing structures of various architectural eras, styles, and structural compositions. Different applications of HBIM technologies are highlighted as well. The BIM industry and the heritage authority or community have grown significantly recently. The evolving HBIM process and tools have brought about a major shift in the sustainable cultural heritage paradigm.

DOI: 10.1201/9781003357483-4

Imagine the added value of digital replicas containing physical and nonphysical information about sophisticated man-made structures for heritage communities. Despite the analysis and evaluation of myriad types of buildings via digital tools, all this leads to the pursuit of sustainability for cultural heritage. A wide range of technological advances has been incorporated to facilitate the process of HBIM, including physical and nonphysical data retrieval. Counsell and Taylor [1] set the requirements for the HBIM framework, which is essentially needed to support the full range from cultural heritage environments to specific structures, including accurate surveyed data via the recording of values and analysis of materials, structure, and pathology. They also highlighted the fact that the use of the structure always reflects environmental, social, cultural, and economic changes through time. The adopted technologies include Laser scanning, BIM modeling tools, Virtual Reality applications, and performance simulation software. There is a wide range of applications in which HBIM Technologies are involved. These include classification, conservation, documentation, retrofit, rehabilitation, and other experiences. It is provided as a sample of these experiences to demonstrate the HBIM schemes within the heritage.

Several research efforts have discussed the classification of heritage components. They provided a classification typology related to heritage components in a historical building [2, 3]. Examples of the varied classes and their classified elements are provided while creating a class evaluation method that emphasizes the architectural elements and suggests proper restoration techniques. Similarly, a general classification is identified to derive a seismic failure plan for each building component. In comparison, finding the relationship between different classification types and types of buildings according to the class of damage and the behavior intensity enabled them to identify the classes according to their seismic activity [4]. Albertini et al. [5] focused on the heritage macro-elements scale and discussed the cultural heritage-related information by handling in a comprehensive, integrated way heritage information is preserved and presented in a soft digital form. Challenges are identified, and a strategy is developed to combine the related heritage information in a clear, comprehensive, accessible soft format for interested parties. The researcher created a portal including the developed strategy of different heritage information control and immersive experience.

Abulnour [6] explained the significance of creating a Monumental Protection Program (MPP), which begins with classifying the elements according to their significance, where rare elements are considered more significant than others. Those elements consist of indoor and outdoor materials. However, conservation requirements should be analyzed to devise a plan for the restoration process, including the deterioration of the building status and the exposed threats. Luccchi [7] emphasized updating the heritage-protecting rules and regulations to protect the monuments effectively. Khodeir et al. [8] reviewed and analyzed world-spread heritage buildings' concerns about their sustainable retrofit. They categorized the Egyptian Heritage and planned to integrate BIM technology in sustainable heritage retrofitting, emphasizing the legislation's significance in supporting the retrofit. The research proposed the cultural classification model. They suggested a stronger heritage classification system in Egypt to facilitate heritage listing and conservation toward a sustainable framework.

In a retrofit study, the significance of artifact protection is indicated to provide visual comfort at recommended illuminance levels for artifacts. They compared the daylight performance of a flat ceiling and a pitched roof ceiling model of a historical museum, revealing the better illuminance distribution of the pitched roof than the flat ceiling [9]. The effect of different skylight design parameters on building performance was studied using a parametric method to provide 100 different design solutions and assess those in Useful Daylight Illuminance (UDI) and Energy Use Intensity (EUI) [10]. Recommendations were provided for the effective skylight-to-floor ratio (SFR) to range within 4–10%, and energy reductions were varied according to building type, glazing type, climate, and light control method [11]. In a regular office ceiling, a skylight opening with a 7% SFR was added with a slated light well to examine and simulate the proposed design to enhance daylighting [12]. Glazing sizing in atrium spaces was discussed, indicating that the most difficult sizes are rectangular-shaped ones. However, the relationship between daylight performance and visual comfort in building space was revealed to aid decision-makers [13].

Experimental and simulation studies have been widely compared to evaluate the accuracy of the utilized simulation software. The literature indicated that simulation methods to assess daylighting in spaces usually provide higher illuminance values than experimental measurements [13, 14]. The Kriging method provided effective daylight prediction that saves annual simulation time. They compared different daylight simulation engines to the prediction method in different sky-cover conditions of the year [15]. Different daylight dynamic and static metrics were criticized and evaluated in assessing human perceived daylight in classroom spaces, where it was found that spatial daylight autonomy (sDA) and Annual Solar Exposure (ASE) are more indicative [14]. The glare problems resulting from large glazing areas incorporated in building spaces were assessed to select the most suitable daylight control element [16]. Previous research indicated the significance of adjusted skylights in increasing daylight adequacy in spaces. However, those lack information on enhancing reused skylights in adapted conditions, and daylight analysis has never been thoroughly analyzed.

It is worth noting that adapted museum design daylight aspects create an additional challenge to minimize lux exposure values [15]. As such, this study provides a comprehensive daylight analysis and guidance for the adapted reuse of similar adapted museum cases located in arid climatic conditions with skylight openings. The study is essential, especially in the case of an old historical space in which anticipation of daylight performance may vary according to improper fitting and material translation in the simulation software. Dimension of the historical challenge extends to the fact that meeting daylight limits are critical to preservation purposes [16] and that historical interior beauty should be reflected with charming daylight. The annual simulation uses a different daylight model and requires time to calculate the varying input values along the varied annual sky conditions.

However, it requires data abstraction, but still, it can reflect real conditions in most cases [13–17].

Chamilothori et al. [18] investigated the applicability of using the VR headset in accurately representing daylight-simulated spaces to occupants in which real and virtual environments are both compared in terms of visual daylight perceptions (pleasantness, interest, excitement, complexity, and satisfaction) and other physical symptoms (eye soreness, tiredness, clearness of vision). The effect of the display method of the virtual environment on occupants is also examined. They ensured success in measuring the user preferences in a virtual environment, and there is a high similarity in their responses to a real environment. Chamilothori et al. [19] demonstrated different daylighting scenes using the grasshopper modeling tool with unity for VR experience to explore the users' satisfaction and suggest the optimum design configuration for façade openings and interior daylighting.

Marín-Morales et al. [20] compared virtual and real museum environments through human navigation tours in an exhibition while proposing questions to illustrate the human experience in both environments. The questions revealed the difference in tour timing. They examined the difference in the sense of space acquisition and compared the heat map visualization of five different exhibition rooms in the physical and virtual environments. Data from the 60 participants were analyzed using their mean value and the standard deviation. This allowed the authors to validate IVES environmental analysis tool through physical existence using HTC Vive, providing a novel validating tool for designers.

Chen et al. [21] studied the applicability of virtual reality in replicating the physical environment and provided an inexpensive light assessing tool compared to scaled models. They used a 360-fly camera to record 360 videos and integrate the panoramic video into a 360-fly App that allows smartphones within a VR headset to visualize the room while moving around in the virtual space. Photo-captured images of the room, the recorded videos, and the VR experience are then compared through the participants' sense of presence and subjective and emotional attributes. They used the average and standard deviation calculation method of the 40 participants' questionnaire findings. Amirkhani et al. [22] aimed to satisfy the occupants of office space in terms of lighting conditions. They investigated the influence of change in the WWR and the electric lighting levels in the room on the occupants' satisfaction in an immersive virtual reality IVR environment. They used Autodesk 3Dmax software to provide the room with virtual materials and rendered scenes. The questionnaire relies mainly on the users' sense of lighting in the space and their satisfaction with the lighting conditions retrieved from four different examined WWRs to assess the rated contrast RC of occupants.

4.3   Reusability Utilizing HBIM

Several research efforts have been conducted to retrofit ancient buildings. Buildings are mainly retrofitted to target comfort and energy efficiency. Case studies target thermal comfort and visual comfort. They promoted minimal intervention into historic fabric such as frames, windows, or adornments on the walls to prevent damage. It is essential to gather data relevant to the studied building, such as historical use, goal, energy use bills, and occupation, to understand the way forward and minimize the physical and visual impact of any work or new equipment. That refers to the need to keep the architectural design as little as possibly changed [23–25]. Ascione et al. [26] proposed a retrofit methodology using field measurements to evaluate the materials' conductivity, energy loads, and others. They used technologies like thermal resistance sensors inside spaces and core sampling analysis. Also, they used a simulated model to perform energy-related analysis, compared the results with those measured data, and then evaluated the overall improvements after adopting the retrofit. De Berardinis et al. [27] provided a methodology for an energy-oriented retrofit historical building. They utilized multiple thermal measurements to evaluate the possible retrofit interventions and their related level of performance, allowing the development of an evaluation matrix aiding the historical building recovery process. The introduced methodology provided retrofit solutions for all the envelope components that enhanced the overall building performance by 50%.

Bellia et al. [28] proposed an energy retrofit plan for a historical building in Italy. With special considerations to the aesthetics and thermal characteristics of the proposed interventions, a feasibility study of Photovoltaic (PV) integration and daylight analysis provides an overview for more convenient integrated strategies from different points of view. The proposed upgrade allowed energy savings of 27.1%. Marzouk et al. [29] provided a framework to identify the theoretical simulated and the actual building performance gap to reach more optimistic real energy-saving figures in terms of the envelope and building retrofit. Bruno et al. [30] investigated an 18th-century noble residence with various diagnostic investigation methods, including photographic survey, cracks mapping, thermography usage, and radar technologies to demonstrate the constructive material properties for efficient building diagnosis. Regarding retrofitting, BIM tools were found useful in providing the elaborative model and reading tables and charts for assessing the building's current properties, as well as in validating the tool for integrated building diagnosis.

Existing buildings are valuable assets to the created industry. It is more sustainable to consider these buildings' improvements, enhancements, and reuse than to build new ones. Existing historical buildings are considered a huge hub of resources and materials that encounter additional value, which should be salvaged, and reused effectively to ensure sustainable reuse of the existing carbon footprint. On top, the Energy crisis in Egypt has been rising in the past few years due to the high increase in energy usage by the building sector. Thus, reusing existing buildings is considered a part of a bigger picture that mainly involves creating sustainable communities with environmental stewardship. Dyson et al. [31] acknowledged the most significant parameters related to any building adaptation. They prioritized retaining the original function whenever possible, adhering to codes and regulations, and managing common risks. However, Rispoli and Organ [32] realized that the adaptive reuse of heritage could be considered the most challenging when it comes to collaboration and communication between different shareholders. Also, they highlighted the significance of skilled experts' involvement in providing energy-saving techniques without damaging the heritage identity. Conejos et al. [33] developed an adaptive reuse framework that accounts for the functional, socioeconomical, physical, technological, and political influence. The framework is tested in two different old heritage buildings in different contexts to assess their possible enhancement potential for future sustainable adaptive reuse, as reframed by the authors in Figure 4.1.

National and international policies have spread awareness of the significance of maintaining, adapting, and reusing heritage. Trends and initiatives have provided guidance and plenty of applications on building and urban scales. These provide lessons to learn and guidance for implementing heritage conservation frameworks and strategies [34–38]. However, these policies are coupled with sustainability practices and climate change adaptation strategies to drive heritage toward an environmental sustainability practice [39–42]. Many studies considered the

Figure 4.1 Adoptive Heritage Reuse Criteria Weights

sustainability assessment approach to analyze the restoration and adaptive reuse capability to meet sustainability goals and promote self-maintained structures of unique entities to the social, economic, environmental, and political views. The studies also provided multiple strategies for selection assessment criteria [35, 36, 40, 43]. However, each heritage structure is a unique case and involves unique measures and values for each criterion; thus, the reusability of heritage must be assessed case by case.

A reusability analysis is performed upon redesign and rebuilding of the palace skylight. A review of the most significant primary and secondary variables is provided in Figure 4.2. The SFR, the roof shape, and the glazing properties are the influential primary factors in the building's daylight and energy performance. Also, the secondary skylight design parameters with great influence on performance are all types of daylight redirecting elements with the most common names, including louvers, shades, diffusers, and so on. A thorough literature review and understanding of the influential factors is followed by creating a parametric model for the skylight and the attached spaces to evaluate the optimum parameters range of design configurations and assess performance. However, genetic optimization is utilized to evaluate redesign criteria and assess energy and daylight performance via Kilowatt per hour kWh, Spatial Daylight Autonomy sDA, and Annual Solar Exposure ASE metrics.

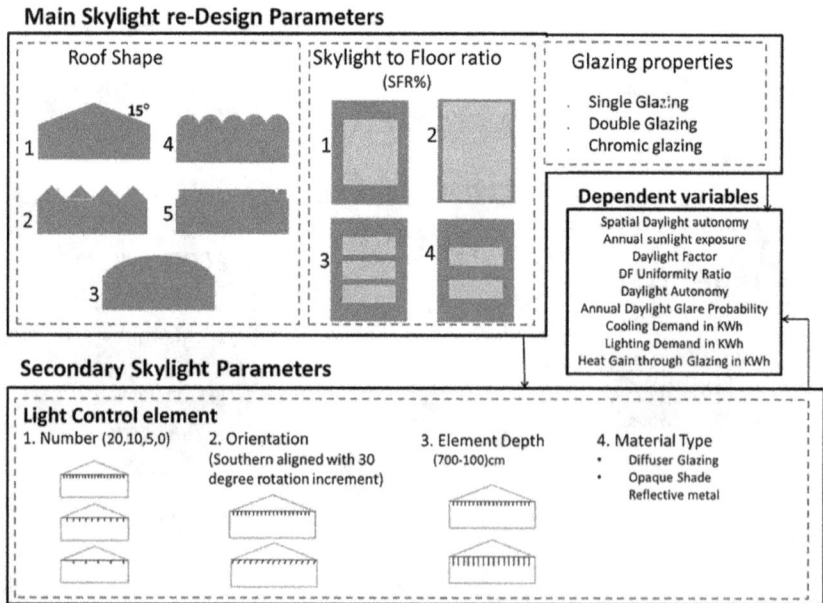

Figure 4.2 Skylight Redesign Parameters

4.4 Proposed Methodology

The main objective is to obtain a historical palace's optimum skylight configuration using qualitative and quantitative analysis methods. The research procedure involves varied tools that achieve visual comfort for potential users of the palace's main hall. The proposed framework consists of three components: the simulation component, the optimization component, and the virtual component. Successive divisions of daylight-related research are followed, as shown in Figure 4.3. Adaptive reuse criteria are utilized to select the most suitable new functionality for the heritage. Subsequently, field measurement values are compared to simulated values to validate the building simulation and provide information for calibration purposes. Simulation results are used to arrive at the most skylight-influencing parameters. Then, optimization is performed to arrive at numerical visual comfort through UDI, kWh, sDA, and ASE variables. Finally, the virtual reality experience gives subjective impressions of potential occupants' visual comfort and experience.

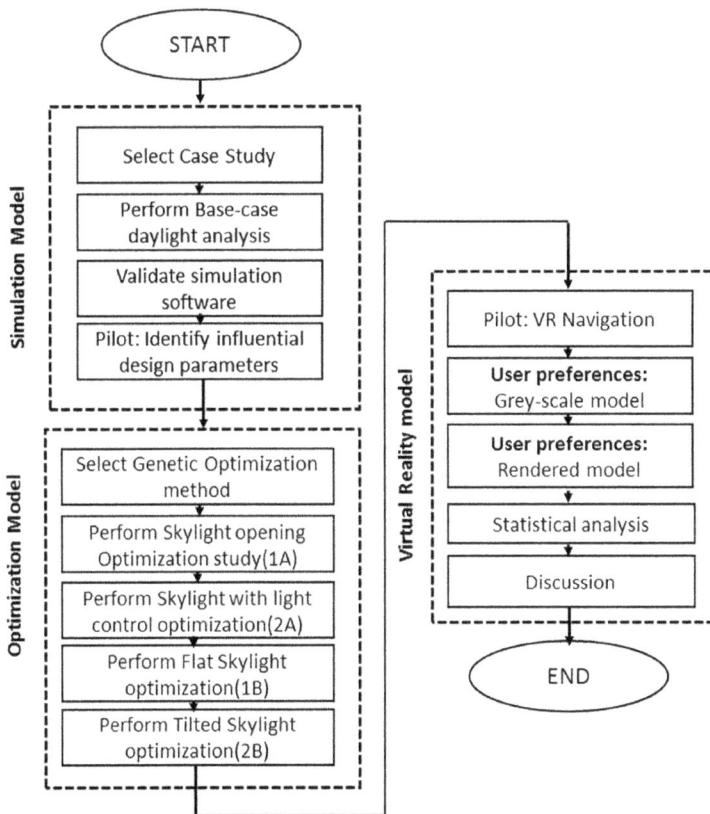

Figure 4.3 Descriptive Study Framework Components

4.5   Case Study

4.5.1   *Case Description*

The case study considers a palace built in the early 19th century for prince Omar Tosson, one of the nobles. It is located in Shubra, Cairo, Egypt. It is considered a highly urban-density zone nowadays. Figure 4.4 shows the location of the Palace in Cairo close to the river Nile. The palace's huge garden is turned into four adjacent school buildings. Physical data is retrieved for the current palace through a laser scanning procedure previously taken and converted to point cloud data, as shown in Figure 4.5. Then, a BIM model is created for the study. In which the study is based on a dynamic simulation of an old historical Palace that represents a typical 19th-century palace shown in Figure 4.6 (Central Hall, Large windows on the facades, high thermal mass, Low occupancy) and incorporates neglected deteriorated status, which is not currently suitable for reuse without a substantial performance enhancement. The model is validated using the same building characteristics as the current Palace state (Floor layout, large stone-bearing wall structure, internal finish, natural ventilation, and density of occupants).

4.5.2   *Case Adaptive Reuse*

Much more palaces and historical buildings are spread all over the world. Not all of them encountered the luck of descent restoration to the built heritage. But even restored ones do not have the luxury of effective adaptive reuse. Using the sustainability assessment approach to analyze the restoration and adaptive reuse capability to meet sustainability goals and promote the self-maintained structure of a unique entity to the social and environmental views. The case study is selected for museum adaptive reuse purposes in the Adaptive reuse component. All necessary assessment criteria are involved to ensure the optimum selection of the new functionality. Selection ensures that economic, environmental, and social concerns are satisfied. The process of reaching the most significant factors contributing to achieving adaptive sustainability included creating a land-use map to know the neighborhood land uses/buildings' functions, then gather all info regarding the constructability/health of the palace built elements, revisit the authority conservation plan. After an assessment of all criteria, one of the most recommended reusability is a museum. Accordingly, the following sections further assess the museum HBIM regarding users' daylight comfort, enhancements, and suitability.

4.5.3   *Running Model Simulation*

All necessary base-case daylight analysis procedures are performed to formulate the main problematic condition within the space, possible potentials, and constraints. A validation study is then performed where simulation data are compared with field measurement readings, which allow calibration values to be identified. That road is the map for identifying the skylight's influential parameters, then

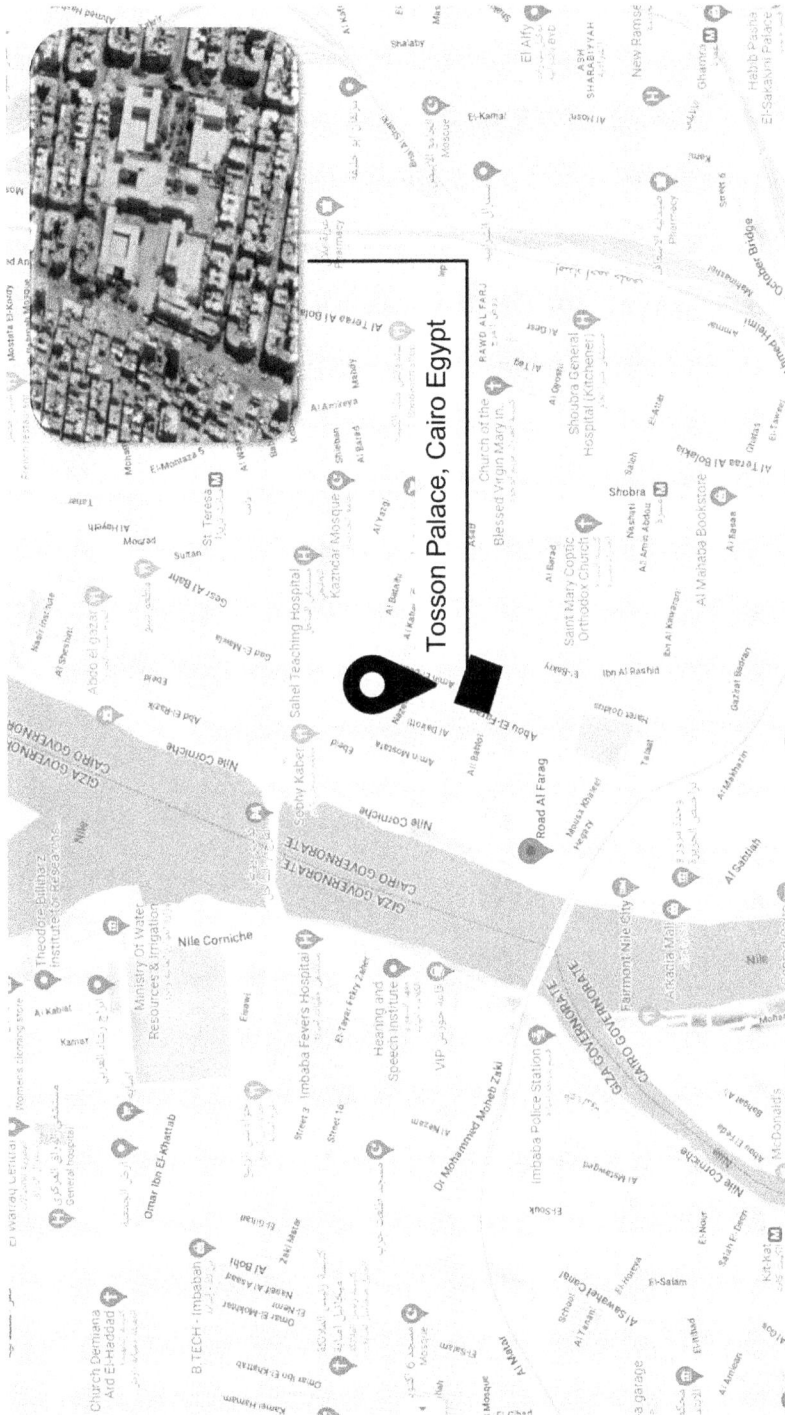

Figure 4.4 Tosson Palace Location Shubra, Cairo, Egypt

Figure 4.5 Tosson Palace Front Façade-Meshed Model from Point Cloud

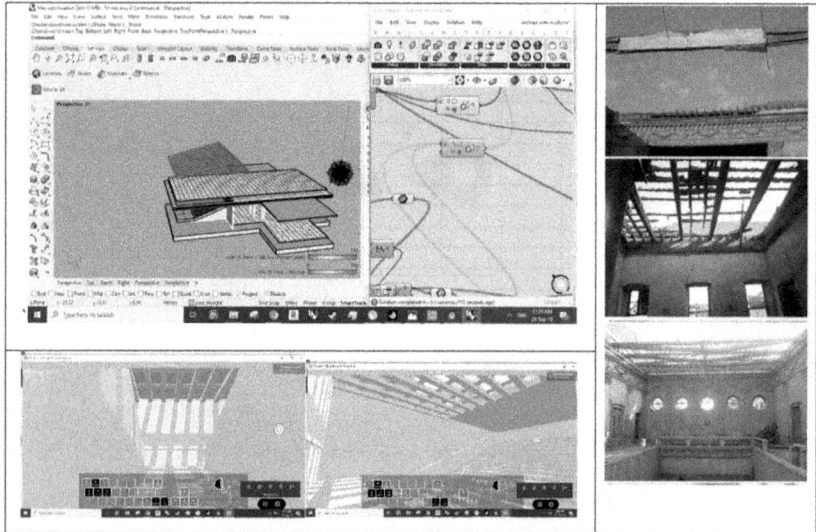

Figure 4.6 Existing Skylight versus Parametric Modeling

simulating their effectiveness in enhancing daylight conditions. Those parameters are further utilized for optimization. Field illuminance measurement is undertaken where measurements are compared to daylight simulation results performed on the same selected day. Lux-meter device was used to provide Lux readings at indicated points. The illuminance was recorded at 60 min intervals from 8:30 to 14:30 during the testing period. The measured direct and diffusion irradiations were used to reflect the initial weather data of the daylighting model in *Daysim*. The experimental and simulation illumination values are taken on the same day with typical weather conditions.

In the validation study, a comparative analysis is performed between the simulated and measured data for the illumination values of the affected points on each floor. The first-floor plan simulated and measured readings are compared. In which the 12 points were distributed on the floor plan area equidistant. The 12 simulation points were both simulated and measured every one-hour interval during a single day. The overall trend of every single assessed point during the day, i.e., from (8:30 to 16:30) is shown in Figures 4.7a and 4.7b to provide a trend analysis for the effect of the skylight on the illumination levels during the day hours. Each curve represents the progress of illumination in a single selected point in different day hours, showing the peak hours of the mid-day at 11:30 and 12:30. The simulated ground floor results in Figure 4.7a are very close to Figure 4.7b; the measured ground floor results except that they are relatively higher in Lux than the measured values and that is justified by the accumulated dirt and dust on the adjacent surfaces. Also, they appear higher in lux, and the lower surfaces' reflectivity also justifies that. In Figure 4.7b, obvious differences between the simulated and measured points highlighted in red and blue, respectively, appear only at points closest to the skylight, those which are exposed to high direct sunlight with maximum lux difference 49% and 24.7% between both assessed values at two different points.

4.5.4   *Running Optimization Model*

Genetic algorithms optimization is utilized to perform multiple optimization problems for the skylight of the main hall. Each optimization problem is formulated to enhance daylight conditions inside the skylight main hall zone. These optimization studies are deeply related and performed simultaneously to draw a corresponding conclusion for the proceeding optimization problem. Several parametric skylight redesign options are tested, as seen in Figure 4.8. The optimum cases of redesign

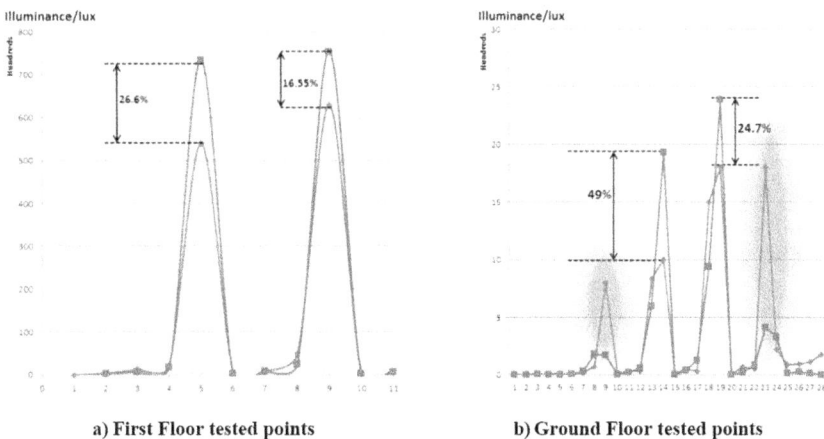

a) **First Floor tested points**   b) **Ground Floor tested points**

Figure 4.7 Simulated Values and Measured Values at 10:30 am (a) Ground Floor (b) First Floor

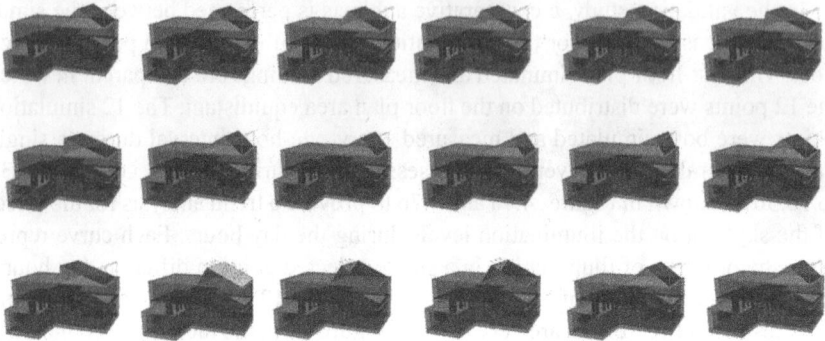

Figure 4.8 Parametric Skylight Configuration

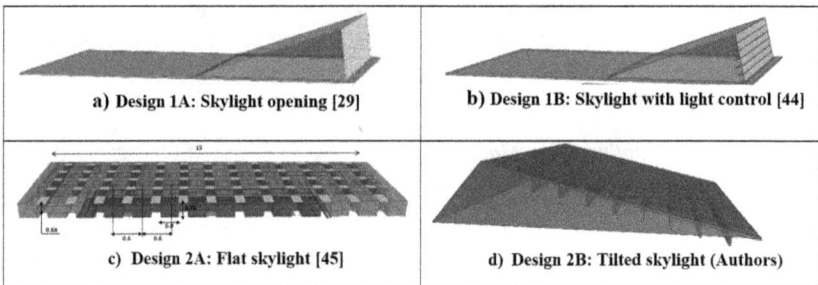

Figure 4.9 Parametric Skylight Designs for Tosson Palace Reusability

represent the most successful skylight redesign options previously studied and suggested by the authors (see Figure 4.9).

Design 2A successfully combines design variables to create an optimized flat skylight. Yet still, there is a clear problem in the effect of 12:00 noon direct daylight on the beneath skylight zone, which is attempted to be solved in Design 2B. The parameters listed in Table 4.1 are utilized with a tilted skylight angle previously suggested as one of the optimum cases in optimization Design 1A. Optimizing the tilted skylight with mullion (Design 2B) provided the most appealing results for daylight uniform distribution with minimum glare values, as depicted in Figure 4.10.

The Pareto-frontier formed by the population points through the optimization process is shown in Figure 4.11. A straight line is formed from the population points, which illustrates a strong, directly inversely relationship within the analyzed sDA and ASE objectives. The shown line is formed from 100 populations after 31 generations. Analyzing the design variables of the optimum cases recommends maximizing mullion spacing to 1.4 and 1.6 in the vertical and horizontal directions,

Table 4.1 Skylight Redesign Parameters

Skylight Glazing Type

Glazing	Single panel	Double panel	Electrochromic	tinted	No of cases
Visible transmittance	88	80	60	30	4

Skylight mullion configuration

Mullion		Interval(m)	Min value(m)	Max value(m)	No of cases
Vertical	V. spacing	0.2	0.6	1.6	6
	V. thickness	0.1	0.1	0.8	8
	V. depth	0.04	0.04	0.8	20
Horizontal	H. spacing	0.2	0.6	1.6	6
	H. thickness	0.1	0.1	0.8	8
	H. depth	0.04	0.04	0.8	20

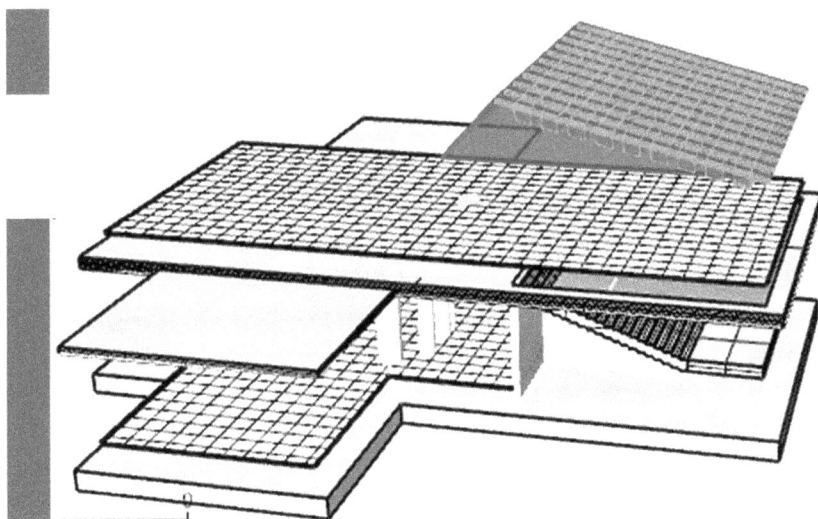

Figure 4.10 Vertical and Horizontal Skylight Mullions Configuration Vary Independently (Located on a Single Side)

respectively. Horizontal mullion thickness is recommended to be thicker than the other perpendicular with a difference of 0.3 m. The recommended depth of the vertical mullion is 0.72 m, and the horizontal mullion is 0.12 m, as shown in Table 4.2.

4.5.5   *Virtual Reality Model*

In the Virtual Reality component, an HTC Vive headset is used to identify the users' preferences for daylight conditions inside the palace's main hall.

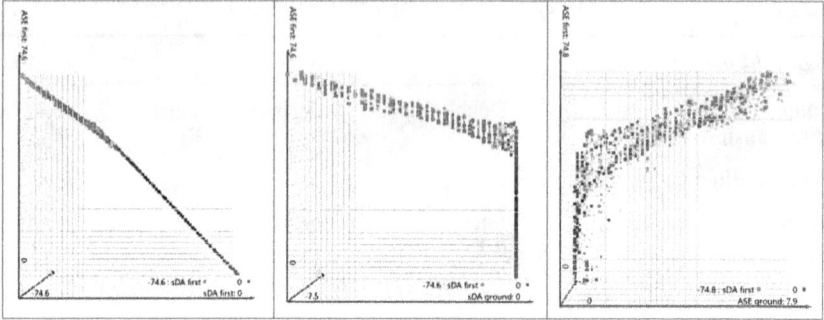

Figure 4.11 Pareto-Frontier Obtained from the Optimization

Table 4.2 Optimum Case sDA and ASE Levels

Mullion Parameters (m.)				Glazing Technology	Objective	
Direction	Thickness	Depth	Spacing	Glazing	sDA%	ASE%
Horizontal	0.5	0.12	1.6	electrochromic	0.1	0.0
Vertical	0.8	0.72	1.4	clear_60	40.6	33.1

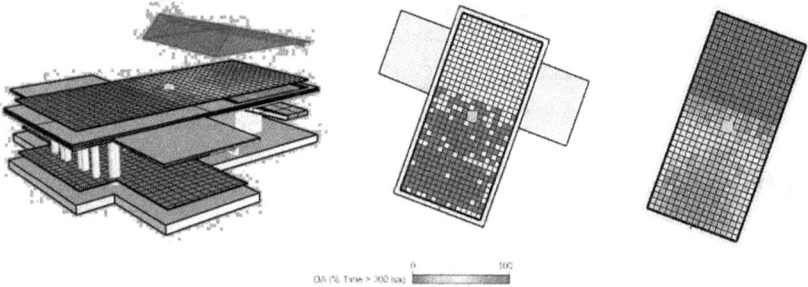

Three introduced skylight designs are compared in a virtual environment where the participants experiencing the VR give their daylight subjective impressions through a post-experiment questionnaire survey. Using the statistical analysis method for the survey data, a logical assessment of their preferences is acquired and evaluated using a five-point scale per Table 4.3, considering six attributes.

The questionnaire contents are divided into the personal information part and scaled measure subjective sense of space attribute part. Participants experimenting were briefed about the experiment, and the three Skylight Designs A, B, and C. Different age groups of participants were considered where they ranged between

Table 4.3 Considered Attributes to Assess Participants' Preferences

	Attribute	Scale				
		1	2	3	4	5
General emotions	**Pleasant**	Very highly pleasant	Highly pleasant	Medium pleasant	Low pleasant	Not pleasant
	Satisfaction	Very highly satisfied	Highly satisfied	Medium satisfied	Low satisfied	Not satisfied
Daylight related	**Contrasting**	Very highly varied	Very highly varied	Medium contrast	Low contrast	Very highly varied
	Brightness	Highly bright	Bright	Medium brightness	Low brightness	Dull/not bright
	Distribution uniformity	High uniformity	Uniform	Medium uniformity	Low uniformity	Non-uniform
	Visual comfort	Very highly comfort	Highly comfort	Medium comfort	Low c omfort	Discomfort

Figure 4.12 Rendered Materials and Wall Hangings are Added

19 and 60 years that covered a wide range of users. The survey experiment data is conducted and collected in three phases. First, the experiment was conducted at a conference related to heritage research studies where most participants were knowledgeable enough and consisted of researchers to professors in the field. The second survey was conducted on architectural students during their class, most of them aged 19 years. The third survey was conducted during a workshop event that gathered practitioners and researchers. For museum reusability, the rendered finishes and furniture added are selected to serve the potential users' functional space, as shown in Figure 4.12. Three Optimum rendered cases are selected based on previous research results [29, 44, 45], then analyzed for user preferences through VR. The overall subjective impression of participants for the three analyzed cases is depicted in Figure 4.13.

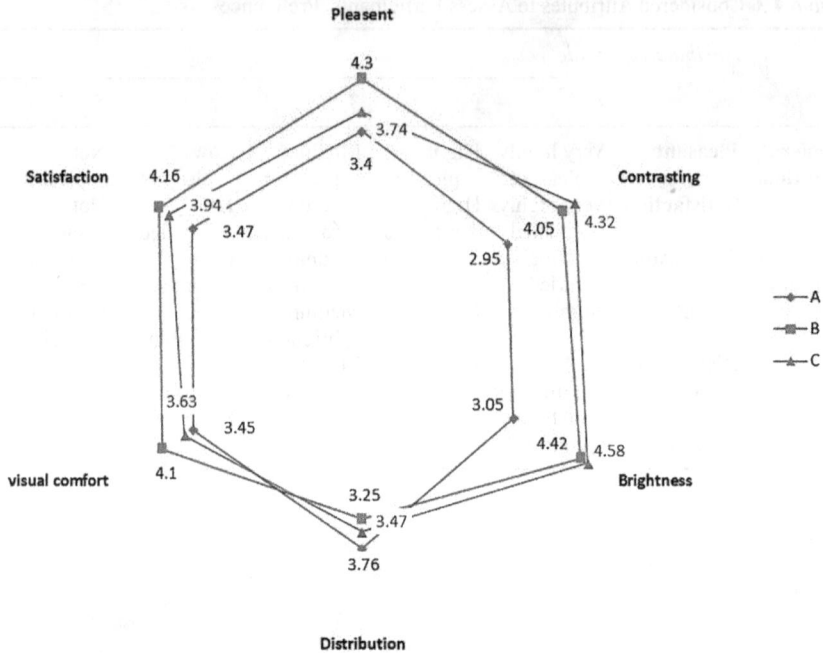

Figure 4.13 Participants' Rating Impressions of Different VR Daylight Experiences

4.6   Conclusion

Heritage Building Information Modeling (HBIM) provides collaboration opportunities for heritage authorities, architects, engineers, and managers. It can systematically and easily implement historic preservation, project planning, and management phases. In addition, the loss of data, documents, and archived issues are typical problems in monument protection. Researchers aim to solve these problems by developing HBIM to intelligently document, interpret, and manage complex and culturally significant historic buildings. A review of these studies has created modern workflows for application areas and processes for HBIM. It describes the tools, methods, and software you may need and potential issues you may encounter while implementing HBIM. The reviewed studies demonstrated digital tools' current applications and possible benefits in achieving a sustainable heritage.

The study is limited to the hot climatic conditions of the selected zone, which is considered the most critical to providing visual comfort in these high daylight intensity conditions. Although simulation results are indicative, under real conditions, some external factors may interfere with the applicability of the results. The difficulty in providing uniform, sufficient daylighting can be rationalized by the hall proportionality, hall form, and uncentered skylight position. Thus, it can be concluded that the skylight variable under study, namely, diffusers' spacing, tilts, material, and skylight inclination, do enhance the daylight quality and quantity in the space; however, in this case, other factors, such as side openings, interior finishes, and ceiling geometry, can be further used for the space to achieve the required daylighting standards.

References

[1] Counsell, J., & Taylor, T. (2014). "What are the goals of HBIM?". In Heritage Building Information Modelling, 2017. (Eds. Arayici, Y., Counsell, J., Mahdjoubi, L., Nagy, G., Hawas, S., & Dweidar, K.). Routledge, New York, pp. 15–31.

[2] Werbrouck, J., Pauwels, P., Bonduel, M., Beetz, J., & Bekers, W. (2020). Scan-to-graph: Semantic enrichment of existing building geometry. *Automation in Construction*, 119, 103286.

[3] Saleeb, N., Garcia, R., & Marzouk, M. (2018, December). "Ontological classification for heritage computer aided design". In 2018 13th International Conference on Computer Engineering and Systems (ICCES). IEEE, Cairo, Egypt, pp. 494–501.

[4] Kloukinas, P., Novelli, V., Kafodya, I., Ngoma, I., Macdonald, J., & Goda, K. (2020). A building classification scheme of housing stock in Malawi for earthquake risk assessment. *Journal of Housing and the Built Environment*, 35(2), 507–537.

[5] Albertini, N., Baldini, J., Dal Pino, A., Lazzari, F., Legnaioli, S., & Barone, V. (2022). PROTEUS: An immersive tool for exploring the world of cultural heritage across space and time scales. *Heritage Science*, 10(1), 1–10.

[6] Abulnour, H. (2013). Protecting the Egyptian monuments: Fundamentals of proficiency. *Alexandria Engineering Journal*, 52(4), 779–785.

[7] Lucchi, E. (2020). Environmental risk management for museums in historic buildings through an innovative approach: A case study of the Pinacoteca di Brera in Milan (Italy). *Sustainability*, 12(12), 5155.

[8] Khodeir, Laila M., Aly, D., & Tarek, S. (2016). Integrating HBIM (Heritage Building Information Modeling) tools in the application of sustainable retrofitting of heritage buildings in Egypt. *Procedia Environmental Sciences*, 34, 258–270.

[9] Ahmad, S.S., Ahmad, N., & Talib, A. (2017). Ceiling geometry and daylighting performance of side lit historical museum galleries under tropical overcast sky condition. *Pertanika Journal of Science and Technology*, 25, 287–298.

[10] Fang, Y., & Cho, S. (2018). "Sensitivity analysis of skylight and clerestory design on energy and daylight performance of a retail building". In 2018 Building Performance Analysis Conference and SimBuild, co-organized by ASHRAE and IBPSA-USA, Chicago, IL, 2, pp. 157–164.

[11] Erlendsson, Ö. (2014). "Daylight optimization-A parametric study of atrium design: Early stage design guidelines of Atria for optimization of daylight Autonomy". Diva Portal.

[12] Ghobad, L., Place,W., & Cho, S. (2012). "Design optimization of square skylights in office buildings". BS2013: 13th Conference of International Building Performance Simulation Association, Chambéry, France, 1, pp. 3653–3660.

[13] Yi, Y.K. (2016). Adaptation of kriging in daylight modeling for energy simulation. *Energy and Buildings*, 111, 479–496.

[14] Li, J., Ban, Q., Chen, X.J., & Yao, J. (2019). Glazing sizing in large atrium buildings: A perspective of balancing daylight quantity and visual comfort. *Energies*, 12(4), 701.

[15] Al-Sallal, K.A., AbouElhamd, A.R., & Dalmouk, M.B. (2018). UAE heritage buildings converted into museums: Evaluation of daylighting effectiveness and potential risks on artifacts and visual comfort. *Energy and Buildings*, 176, 333–359.

[16] IESNA, I. (2012). "LM-83-12 IES Spatial Daylight Autonomy (sDA) and Annual Sunlight Exposure (ASE)". IESNA Lighting Measurement, New York, NY.

[17] Bauer, C., & Wittkopf, S. (2015). Annual daylight simulations with EvalDRC–Assessing the performance of daylight redirection components. *Journal of Facade Design and Engineering*, 3(3–4), 253–272.

[18] Chamilothori, K., Chinazzo, G., & de Matos Rodrigues, J.P., et al. (2018). Perceived interest and heart rate response to façade and daylightpatterns in Virtual Reality, In: Proceedings of Academy of Neuroscience for Architecture (ANFA2018), La Jolla, CA.

[19] Chamilothori, K., Wienold, J., & Andersen, M. (2019). Adequacy of immersivevirtual reality for the perception of daylit spaces: Comparison ofreal and virtual environments. *Leukos*, 15, 203–226.

[20] Marín-Morales, J., Higuer a-Trujillo, J.L., & De-Juan-ripoll, C., et al. (2019). Navigation comparison between a real and a virtual museum: Time-dependent differences using a head mounted display. *Interacting With Computers*, 31, 208–220.

[21] Chen, Y., Cui, Z., & Hao, L. (2019). Virtual reality in lighting research: Comparing physical and virtual lighting environments. *Lighting Research & Technology*, 51(6), 820–837.

[22] Amirkhani, M., Garcia-Hansen, V., Isoardi, G., & Allan, A. (2018). Innovative window design strategy to reduce negative lighting interventions in office buildings. *Energy and Buildings*, 179, 253–263.

[23] Apollonio, F.I., Gaiani, M., Fallavollita, F., Ballabeni, M., & Zheng, S. (2014, November). "Bologna Porticoes project: 3D reality-based models for the management of a wide-spread architectural heritage site". In Euro-Mediterranean Conference. Springer, Cham, pp. 499–506.

[24] Yılmaz, B.Ç., & Yılmaz, Y. (2022). Re-considering the energy efficient retrofitting approach to question cost-optimality and nZEB under COVID-19 measures. *Building and Environment*, 219, 109227.

[25] Hegazi, Y.S., Shalaby, H.A., & Mohamed, M.A. (2021). Adaptive reuse decisions for historic buildings in relation to energy efficiency and thermal comfort—Cairo citadel, a case study from egypt. *Sustainability*, 13(19), 10531.

[26] Ascione, F., Bellia, L., & Minichiello, F. (2011). Earth-to-air heat exchangers for Italian climates. *Renewable Energy*, 36(8), 2177–2188.

[27] De Berardinis, P., Rotilio, M., & Capannolo, L. (2017). Energy and sustainable strategies in the Renovation of Existing Buildings: An italian case study. *Sustainability*, 9(8), 1472.

[28] Bellia, L., Alfano, F.R.D.A., Giordano, J., Ianniello, E., & Riccio, G. (2015). Energy requalification of a historical building: A case study. *Energy and Buildings*, 95, 184–189.

[29] Marzouk, M., ElSharkawy, M., & Eissa, A. (2020). Optimizing thermal and visual efficiency using parametric configuration of skylights in heritage buildings. *Journal of Building Engineering*, 31, 101385.

[30] Bruno, S., De Fino, M., & Fatiguso,F. (2017). Historic building information modeling towards building diagnostic data management. A case study. *TEMA: Technologies, Engineering, Materials and Architecture*, 3(2), 99–110.

[31] Dyson, K., Matthews, J., & Love, P. (2016). Critical success factors of adapting heritage buildings: An exploratory study. *Built Environment Project and Asset Management*,6(1), 44–57.

[32] Rispoli, M., & Organ, S. (2019). The drivers and challenges of improving the energy efficiency performance of listed pre-1919 housing. *International Journal of Building Pathology and Adaptation*,37(3), 288–305.

[33] Conejos, S., Chew, M.Y., & Yung, E.H. (2017). The future adaptively of nineteenth century heritage buildings. *International Journal of Building Pathology and Adaptation*, 35(4), 332–347.

[34] Shehata, A.M. (2022). Current trends in urban heritage conservation: Medieval historic arab city centers. *Sustainability*, 14(2), 607

[35] Haroun, H.A.A.F., Bakr, A.F., & Hasan, A.E.S. (2019). Multi-criteria decision making for adaptive reuse of heritage buildings: Aziza Fahmy Palace, Alexandria, Egypt. *Alexandria Engineering Journal*, 58(2), 467–478.

[36] Othman, A.E., & Mahmoud, N.A. (2022). Public-private partnerships as an approach for alleviating risks associated with adaptive reuse of heritage buildings in Egypt. *International Journal of Construction Management*, 22(9), 1713–1735.

[37] Mısırlısoy, D., & Günçe, K. (2016). Adaptive reuse strategies for heritage buildings: A holistic approach. *Sustainable Cities and Society*, 26, 91–98.

[38] Günçe, K., & Mısırlısoy, D. (2019). Assessment of adaptive reuse practices through user experiences: Traditional houses in the walled city of Nicosia. *Sustainability*, 11(2), 540.

[39] Tu, H.M. (2020). The attractiveness of adaptive heritage reuse: A theoretical framework. *Sustainability*, 12(6), 2372.

[40] Chen, C.S., Chiu, Y.H., & Tsai, L. (2018). Evaluating the adaptive reuse of historic buildings through multicriteria decision-making. *Habitat International*, 81, 12–23.

[41] Foster, G. (2020). Circular economy strategies for adaptive reuse of cultural heritage buildings to reduce environmental impacts. *Resources, Conservation and Recycling*, 152, 104507.

[42] Yazdani Mehr, S. (2019). Analysis of 19th and 20th century conservation key theories in relation to contemporary adaptive reuse of heritage buildings. *Heritage*, 2(1), 920–937.

[43] Bottero, M., D'Alpaos, C., & Oppio, A. (2019). Ranking of adaptive reuse strategies for abandoned industrial heritage in vulnerable contexts: A multiple criteria decision aiding approach. *Sustainability*, 11(3), 785.

[44] Marzouk, M., Eissa, A., & ElSharkawy, M. (2020). Influence of light redirecting control element on daylight performance: A case of Egyptian heritage palace skylight. *Journal of Building Engineering*, 31, 101309.

[45] Marzouk, M., ElSharkawy, M., & Mahmoud, A. (2022). Optimizing daylight utilization of flat skylights in heritage buildings. *Journal of Advanced Research*, 37, 133–145.

5 Digital Documentation of Heritage Buildings

Mahmoud Metawie

Assistant Professor, Structural Engineering Department, Faculty of Engineering, Cairo University, Postal Code 12613, Giza, Egypt

Mohamed Marzouk

Professor of Construction Engineering and Management, Structural Engineering Department, Faculty of Engineering, Cairo University, Postal Code 12613, Giza, Egypt

5.1 General

Heritage documentation is the primary step for any heritage preservation or rehabilitation efforts. Digital documentation can be considered adequate and essential in any heritage rehabilitation project. In Egypt, the current practice mainly depends on 2D documentation techniques, with the data being acquired with traditional survey works and photogrammetry. This current documentation and data acquisition approach is time-consuming, labor-intensive, and prone to human error. Moreover, it is not practically adequate to capture all the fine details and imperfections of the antiquities. WAR, terrorism, unplanned human urban activities, or even unintended acts of negligence endanger heritage sites and put the current documentation efforts in a lagging position behind the deterioration and loss of heritage sites and antiquities. To overcome such lag, this research explores the use of LIDAR and HBIM technologies through a flexible framework suitable for Egyptian Heritage Documentation efforts.

5.2 Introduction

Digital documenting of heritage includes documenting heritage artifacts, places, and even whole landscapes in their present state. The selection of the appropriate techniques for undertaking a digital documentation scheme depends mainly on the size and scale of the documented object and the planned purpose of the dataset. Digital documentation techniques differ according to the scale of the documented subject, whether it is a small object, large structure, or landscape environment. This chapter emphasized the use of virtual reality (VR), photogrammetry, and terrestrial laser scanning (TLS) as methods for digitally documenting heritage buildings. Terrestrial laser scanning TLS is one of the main techniques in the field of digital recording of natural and built environments. It enables the quick and accurate collection of a high-resolution 3D dataset. This is increasingly important for specific

DOI: 10.1201/9781003357483-5

survey types where field scanning time can be reduced, and the risks associated with obtaining measured data may be reduced in difficult access areas. These systems are normally tripod-mounted systems, which have established data capture workflows. The use of TLS for heritage documentation has become increasingly popular. Terrestrial laser scanners have been used for documenting large historical monuments or sites, such as the project on Forum of Pompei [1], Great Buddha projects [2], Romanian Heritage Monuments [3], and Nasif Historical House in Jeddah [4].

Photogrammetry is a wide range of techniques used in photography to obtain 3D information due to differences in line of sight between two or more pictures of a single object. Recent developments in computer hardware and computer vision have led to photogrammetry development. It quickly became a widely adopted and effective technique for generating highly accurate 3D data. It works by simultaneously reconstructing the scene's geometry and the camera's position and orientation in a bundle adjustment procedure, using a range of features extracted from overlapping images. While previous techniques needed a calibrated lens and camera and configuration with known variables, these can now be automatically calculated during alignment [5].

VR is considered a technology that acts as a natural extension in the 3D computer environment associated with advanced inputs and output devices [6]. In other words, VR is defined as an interactive computer-generated environment with three-dimensional objects and locations that can simulate both planned/designed models and real-world scenes [7]. Successful measures and case studies have been conducted to reconstruct demolished or existing heritage buildings or sites virtually. Fernández-Palacios et al. [8] have introduced VR devices to present an immersive experience for 3D reconstructed historical environments. They integrated Oculus Rift and Kinect, aiming to interact with and navigate digitally reconstructed archaeological sites. This also enables viewers to access, visit, and explore virtual environments of heritage sites that have limited access due to conservation strategies.

Ramsey [9] introduced a project that intended to create an immersive 360° experience of the historic city of Wolverhampton. It allows the user or viewer to experience how the city of Wolverhampton may seem in the past from a person's perspective. He has concluded that despite the full details and definite accuracy of a virtual environment, it is not similar to the real environment. It is supposed to be considered an experience on its own rather than a replacement for the real environment. The 3D reconstruction of heritage sites cannot be an alternative to the heritage conservation of existing assets.

Deggim et al. [10] have presented the development of a 3D reconstruction of the two historical cities, "Segeberg" and "Gieschenhagen." Currently, it is called "Bad Segeberg" that is located in Schleswig-Holstein, in Germany. It goes back to the Early Modern Age, around the year 1600. They have accomplished the VR application Segeberg 1600, including numerous historical buildings, major structures, and surrounding sceneries. The handheld controllers enable the viewer to navigate through the virtual environment.

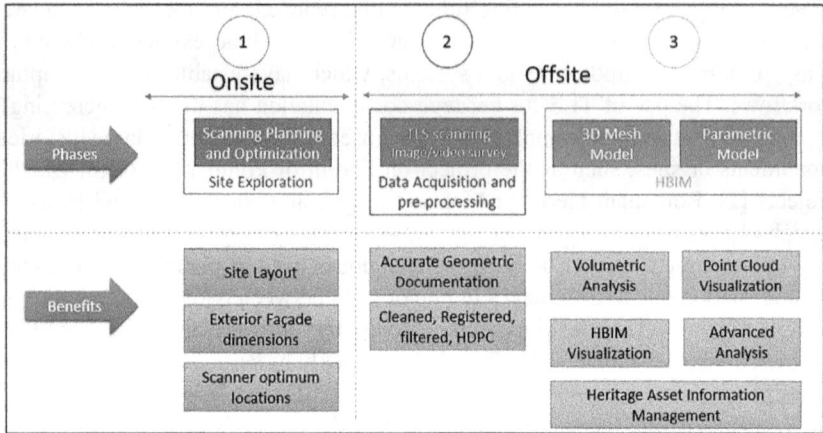

Figure 5.1 Schematic Diagram of the Proposed Framework

5.3   Developed Framework

LIDAR, photogrammetry, and VR technologies have great potential in documenting and visualizing heritage sites. This research proposes a framework for planning heritage building rehabilitation using digital documentation and visualization. The framework overcomes the weaknesses in the traditional documentation methods, which rely heavily on data acquisition techniques that require time and effort, such as manual measurements, survey points, or traditional photogrammetry. LIDAR data is the proposed framework's pivot to overcome the traditional methods' weaknesses. The process of applying the proposed framework is divided into three phases as follows (see Figure 5.1):

- Phase 1: Site Exploration and scanning planning optimization
- Phase 2: Data acquisition and preprocessing
- Phase 3: HBIM modeling for planning rehabilitation.

5.4   Site Exploration and Scanning Planning

This phase is divided into two stages: the first is site exploration, and the second is planning and optimizing the scanner's optimum locations. One of the first tasks of the site exploration phase is making an investigation tour of the heritage site. In this tour, simple sketches, notes, videos, pictures, and initial measurements can be integrated to have a complete plan for the site. Also, a total station survey or initial TLS scanning could be used in this stage. All the gathered information contributes to the final decision regarding the type of digitization equipment, the considered safety issues, the equipment locations, and the position constraints. Also, the different onsite equipment, the crew size, and the equipment handover and maneuvering could be determined in this phase. The site's layout, the surrounding terrain, and the building materials are considered when planning the optimum viewpoints' locations in the next stage. Some heritage sites have restrictions to access certain areas and allowable access time. These restrictions have a strong effect on the data acquisition phase. Also, it is very important to ensure an electricity source's availability to charge the intended used equipment. Some digitization techniques, such as LIDAR and photogrammetry, may require placing special targets, scale bares, or spheres before the acquisition phases, as shown in Figure 5.2. These objects are very important in the preprocessing stage to ensure accurate registration and geo-referencing for a large-scale project's gathered data. In addition, the team has to confirm the possibility of using stable cranes, building temporal scaffoldings, or accessing nearby buildings where data accusation can be performed.

The planning stage starts after completing the site exploration stage. All information gathered in the previous stage is used in the planning task to determine the number of scans required to capture the features of the target object from all directions while maintaining an adequate level of accuracy. The planning stage focuses on scanning object geometry factors to increase the quality of the generated point cloud of the heritage buildings. It optimizes the scanner locations and field of view to increase the point cloud quality and shorten the scanning time while

Figure 5.2 Some Special Targets for Different TLS Equipment

guaranteeing a set of quality constraints for the point cloud. The quality constraints are based on the incidence angle between the scanned surface and the laser ray and the max spacing between points. Three multiobjective optimization algorithms are utilized: (1) genetic algorithm, (2) Jaya algorithm, and (3) particle swarm optimization to increase the quality. Two optimization performance measures are adopted to compare the outputs of the optimization algorithms. Finally, a multicriteria decision-making technique (Weighed sum model) is used to choose the optimum solution between the Pareto frontier solutions. Optimization algorithms minimize point cloud density and scanning time while assuring a required point spacing and max incidence angle by changing the distance between the laser scanner and scanned facade, horizontal and vertical scan repetitions, and scanner different resolutions.

5.5   Data Acquisition and Processing

This phase is divided into two steps. The first step is using a TLS scanner and a digital camera to scan and capture the texture of the heritage building according to the information gathered in the previous phase. The second step is processing the laser scanning data and the photography data. The first step results in accurate documentation of the geometrical and texture features of the site. In contrast, the second step results in a clean registered point cloud which can be solely used in visualization and even the production of 2D plans for legacy systems. Also, it results in accurate texture documentation that will be used for texturing the generated 3D models. Also, mixing TLS and photogrammetry data acquires a greater degree of site coverage and eliminates possible shadows.

After completing the planning and optimization phase, the terrestrial laser scanner scans all the building facades and interior rooms. Usually, terrestrial laser scanners are tripod-mounted systems (see Figure 5.3) that capture reality with a conventional workflow and processing steps. The output data are called point clouds, a collection of millions of discrete points per scan (see Figure 5.4). Each individual point has seven different records: three values for the XYZ coordinates, three for the RGB, and one for the intensity. The intensity value is the energy strength of the laser ray when it is returned to the scanner. It depends on many factors, such as distance traveled and the scanned object's materials.

During TLS scanning, a digital camera captures all the textures and features of the heritage. The quality of the photographs taken during the data acquisition process determines the quality of the 3D model and point cloud produced via the photogrammetry process. Using photogrammetry for site-scale capture involves thoroughly capturing all accessible and visible surfaces from the greatest number of unique angles. In practical terms, taking photographs directed at the subject from near-ground level, standing height, and elevated position should give stronger vertical coverage. Repeating this incrementally at offset around the perimeter of the surfaces should give an even overall coverage (see Figure 5.5).

Photogrammetry produces 3D spatial data in the form of a point cloud and 3D model and enables the creation of texture maps of the surface using the captured images. Usually, the photogrammetry process depends on automated process steps

Figure 5.3 Terrestrial Laser Scanner at the Main Entrance of Omar Tosson Palace

Figure 5.4 The Resulting Point Cloud at the Main Entrance of Omar Tosson Palace

that are computationally intensive operations on the system CPU (Central Processing Unit) and GPU (Graphics Processing Unit) while using large amounts of system memory. Combining datasets from LIDAR and Photogrammetry sources is necessary to acquire a greater degree of site texture with an accurate scale. The second stage is processing the LIDAR point cloud and the digital camera images. TLS scanning processing consists of several actions, including cleaning, registration,

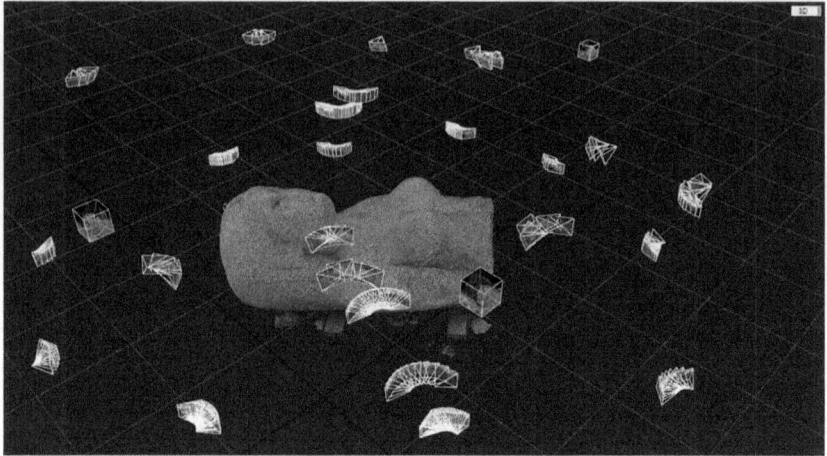

Figure 5.5 Aligned Camera Positions for the Created 3D Model

and filtering point cloud, while photogrammetry processing is an automated process with highly computationally intensive operations.

5.6   HBIM Modeling

HBIM modeling consists of several steps, such as parametric modeling for normal heritage elements, nonparametric modeling of complex heritage elements such as complex roofs or domes, and creating and adding heritage parameters to the created models to form a full Heritage asset information model (HAIM). The modeling of the HBIM model starts once the 3D point cloud has been processed, and it consists of the following:

1 Modeling main body skeleton
2 Creating families for regular structural and architectural elements
3 Creating families for irregular structural and architectural elements
4 Inserting the created library in the 3D model saves considerable time and work.
5 Adding Heritage parameters to all families and details
6 Visualizing the HBIM model using VR

5.6.1   *Modelling Main Body Skeleton*

Heritage building's main skeleton is modeled using the point cloud data from the previous stages. After completing the scanning and photogrammetry data processing for the heritage site and generating a clean processed textured point cloud, the point cloud is inserted into a BIM software for reverse engineering. Autodesk Revit is used as it offers quick modeling, 3D model adjustments, and high-standard documentation, as well as a high level of flexibility. Also, Leica CloudWorks (a Revit plugin) is used to help in reverse modeling. CloudWorks is used for importing and clipping point cloud into a Revit project, as shown in Figure 5.6. Modeling begins with the clipping point cloud, which separates the floors to facilitate the modeling of the floors themselves and the internal and external walls of the building (see Figure 5.7).

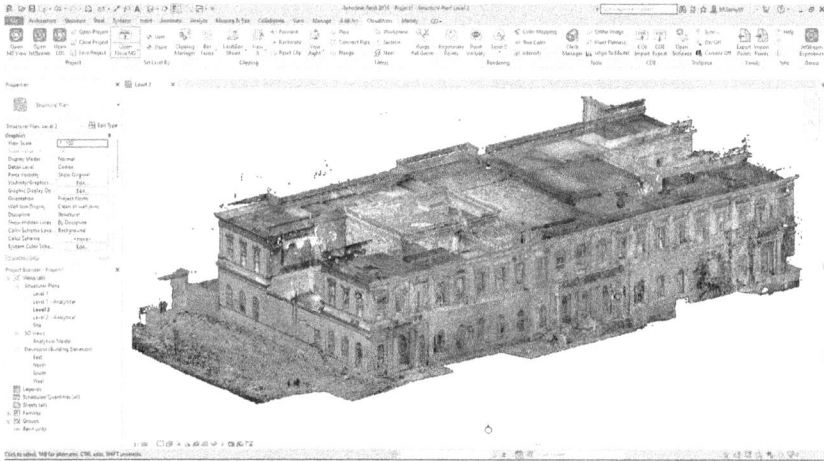

Figure 5.6 Imported Point Cloud in Revit

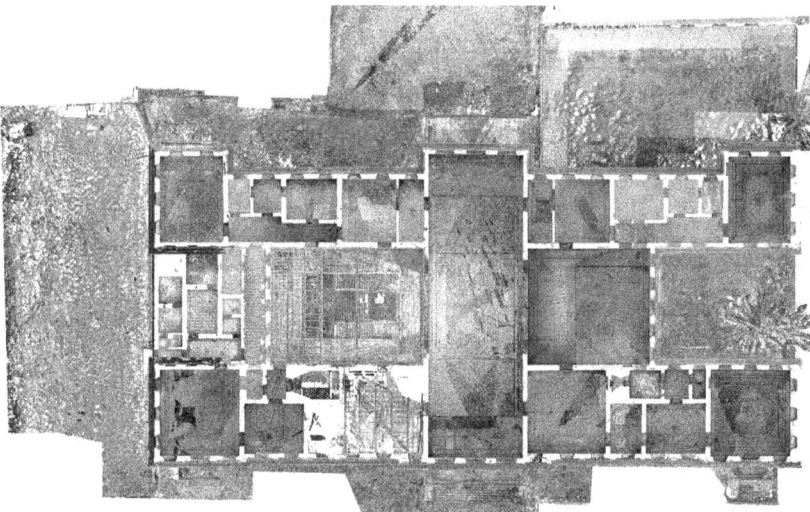

Figure 5.7 Point Cloud Plan of the First Floor

5.6.2   *Modeling Repeated Regular Heritage Elements*

Heritage building elements can be categorized as regular and irregular elements. Regular elements could be modeled using normal BIM software; however, it would be impossible to model irregular elements using BIM software. After completing skeleton modeling at the previous stage, laser scanning and imaging data are used to generate HBIM families for the regular heritage elements. This approach can be considered a reverse engineering procedure where an object's geometry, physical dimensions, and material attributes are collected to create HBIM families. This step aims at reducing heritage building complexity by separating and modeling the repeated architectural or structural elements from the building skeleton. Regular elements point cloud is separated from the whole point cloud and imported in the BIM software. Figure 5.8 shows the point cloud of a Heritage Window inserted in a BIM Software with the dimensions placed on it. This facilitates building the HBIM family, as shown in Figure 5.9.

Figure 5.8 Point Cloud for Architectural Window

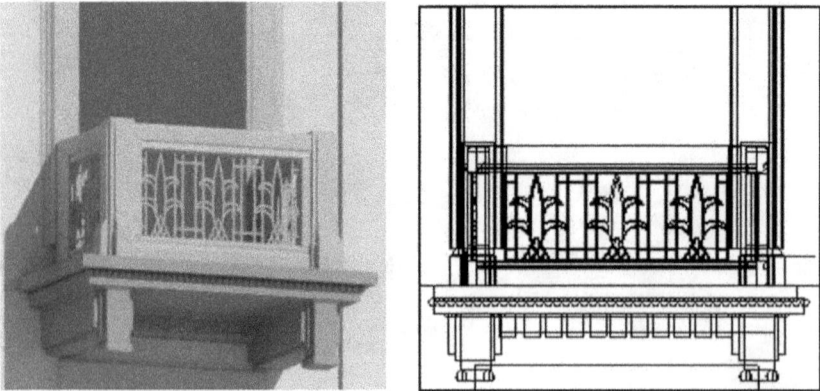

Figure 5.9 HBIM Family for Architectural Window

5.6.3 Modeling Irregular Heritage Elements Families

Irregular Heritage elements cannot be modeled using standard BIM software, so the point cloud of these elements is meshed and then imported into the BIM software as an irregular family. The meshing process is used for complex irregular elements shape that cannot be modeled using BIM software, such as irregular domes or artifacts on windows and facades (see Figure 5.10).

The point cloud of the irregular elements is segmented and separated from the whole point cloud, which is subjected to cleaning from noise to accelerate the exportation and ease the transformation process (see Figure 5.10). Reality capture software [11] integrates both the laser scanning data and the imagery data. It is a photogrammetry software that creates 3D models of unordered images and terrestrial laser scans. Combining datasets from laser scanning and photogrammetry is crucial to get an accurate scaled textured 3D model. The point cloud is imported in Reality Capture as an E57 file format, while photogrammetry images are imported as JPG files. The generated point cloud is then converted to a polygon mesh. Converting a point cloud to a polygon object (commonly called wrapping) is like stretching plastic sheeting around a point object and pulling it tight to reveal a polygon surface. It is advantageous to wrap a point object because many refinements can be performed on a polygon object that cannot be performed on a point object. Figure 5.11 shows the meshed point cloud for the previous elements shown in Figure 5.10. Finally, all created elements are inserted in the skeleton model. This saves considerable time and work and reduces HBIM file size.

Figure 5.10 Separated Point Cloud for Different W Irregular Elements

Figure 5.11 3D Meshing for the Irregular Elements

5.6.4　Adding Heritage Parameters

BIM can be described in simple terms as a digital process to visualize all the elements of a building. In contrast, it is described as object-orientated parametric modeling in scientific terms. The BIM process includes assembling smart objects into a digital representation of a facility or building (see Figure 5.12). These objects contain geometric data (2D and/or 3D) and nongeometric related data. They are parametric, identified by rules, and modified in their context automatically. Data are incorporated into the model by applying the relevant information to the corresponding BIM objects. Thus, BIM is a digital information platform for the asset created.

BIM objects are intelligent parametric components. They include both geometry (2Dimensional or 3Dimensional) and related information. The object description is based on a group of parameters that describe the object's behavior. An HBIM model contains the following:

• Geometry information,
• Nongeometric data, and
• Linked sheets and documents.

The geometric model data are displayed in the 3D views, drawing sheets, and schedules within the BIM environment. Model views are synchronized, i.e., all views are changed continuously to reflect model changes. It means that the outcomes of the model details (in the form of schedules, drawings, and visualizations) are still consistent.

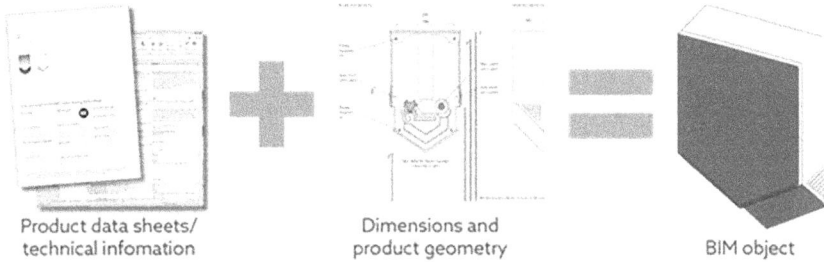

Figure 5.12 BIM Object Contents

Nongeometric information may include the physical features of the created object, such as appearance, materials, and condition. It may include operations and maintenance details such as product and model names, guarantee period, mainte-nance and repair orders, training specifications, documentation, inspection dates, replacement costs, and requirements for H&E. Nongeometric information may also include structural, environmental, and mechanical efficiencies, such as dis-posal instruction, power consumption, recycling instructions, standard compliance, and load-bearing capacity. In heritage, information such as historical, cultural, and architectural values, age, and significance should be part of the model and attached to the individual components or rooms or the building itself (see Figure 5.13). The information parameters distinguishing HBIM from conventional BIM are conser-vation parameters that are gathered and connected to the HBIM model. Heritage parameters are required for heritage rehabilitation planning and the management of heritage assets and are established and categorized under different criteria, includ-ing facility, element, and Room or space parameters.

Table 5.1 lists a set of created parameters that can be used as a guide for any her-itage project. Nevertheless, each corporation can build its own parameters depend-ent on the project's needs. An HBIM project template is created and filled with these parameters for any heritage project.

5.6.5   VR Using the HBIM Model

After filling the HBIM model with the heritage data, VR could be used to visualize this model for more collaboration and dissemination. The HBIM model should be optimized for VR as following steps:

1   **Create a 3D view that will be visualized using VR**: Only the geometry pre-sented in the created view at the time of exportation is exported.
2   **Hide unnecessary categories**: Anything modeled or connected to the 3D view may not be necessary to be visualized using VR. Repeated Categories such as nuts, rebars, or bolts should be hidden as the best and simplest way to enhance the VR environment without losing details geometry, as shown in Figure 5.14.

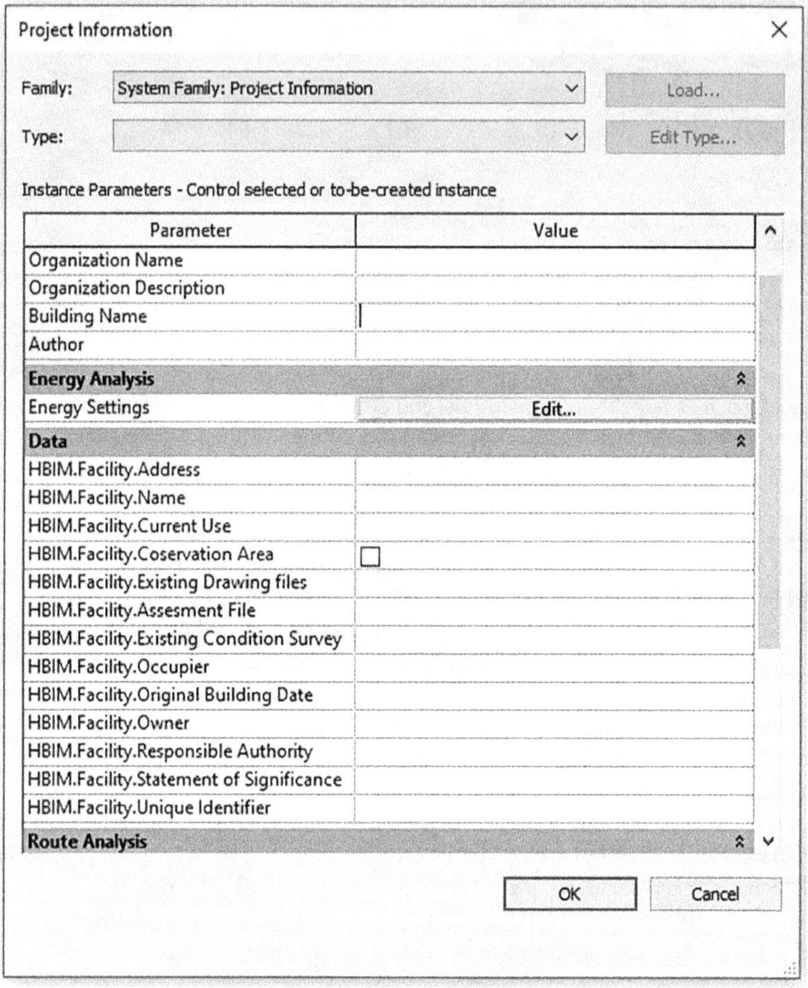

Figure 5.13 Example of Facility Parameter Created for Heritage Building

3 **Hide Linked files and worksets**: They should be hidden and removed from the VR view as they increase the model size and affect VR performance badly (see Figures 5.15 and 5.16).
4 **Define the geometric bounds**: By default, the boundaries of the visualized model are the whole model extent. These boundaries should be changed to view the desired VR view. Section box in Revit could show only the desired geometry from the VR view (see Figure 5.17). This will also enhance the VR experience and make it smoother.

Table 5.1 Proposed Heritage Parameters for Data Exchange

Facility Parameters	Element Parameters	Room/Space Parameters
• Asset number/unique identifier • Occupier • Owner • Responsible authority/trust manager/etc. • Building/monument name • Building/monument address • Conservation designation (grading) • Scheduled monument number/list entry number (link to listing description) • National grid reference • Conservation area (yes/no) • Property type (heritage estate – roofed, un-roofed, ruin, etc.) • Current status/use • Original building date • Statement of significance/ heritage assessment (link to document) • Conservation management plan (link to document) • Existing drawings (link to documents) • Existing condition survey (link to document) • Asbestos (Y/N) – (link to asbestos register/ management plan) • Structural integrity (Y/N) – (link to structural survey) • Project files (archived – link to document) • O&M manual (link to document) • Ecology issues (Y/N) – (link to ecology survey) • Photograph – current/ historic (link to documents) • Archive Records – (link to the list of archive references) • Insurance (Y/N) (Link to document) • Laser scanning survey	• Unique identifier (UID) • Historic/common name (historical/architectural features) • Year of construction/ historical construction phase • Installation date • Life expectancy • Original/historic fabric (Y/N) • Level of significance (value) • Material – high level (stone, timber, etc.) • Material specification (link to stone matching report/ petrographic analysis, etc.) • Color • Finish/mix/ratio (link to mortar analysis, etc.) • Material source/supplier • Sustainable material (Y/N)/ sustainability note • Asbestos (Y/N) • Condition – good, fair, poor, etc. • Priority – urgent, within 12 months, within 24 months, etc. • Defect description • Current defect (Y/N) • Repair/task • Reversible repair (Y/N) • Cost/defect liability • PPM task/cyclical maintenance task (planned and preventative maintenance) • Ecology issues (Y/N) – bats, bees, birds, newts, etc. • Environmental monitoring (Y/N) – (link to doc/reading) • Existing drawings (Y/N) – (link to drawings) • Archaeological record (Y/N) – (link to document) • Other specialist reports (Y/N) – (link to document) • Facility name • Room name	• Room number • Room name • Known as – historic name • Use • Space classification/ type • Room/space significance • Room/space vulnerability • Access details (public access Y/N) • Historical collections/ interiors (Y/N) • Perimeter • Height • Floor area • Volume • Survey documentation (link to document) • Survey notes • Structural integrity (Y/N) – (link to structural survey) • Historical documentation (link to document – room datasheet) • Historical notes • Asbestos (Y/N) (link to asbestos register) • Asbestos notes • Ecology issues (Y/N) – (link to ecology survey) • Photograph – current/ historic (link to documents) • Environmental monitoring (Y/N) – (link to documents or readings) • Structural System

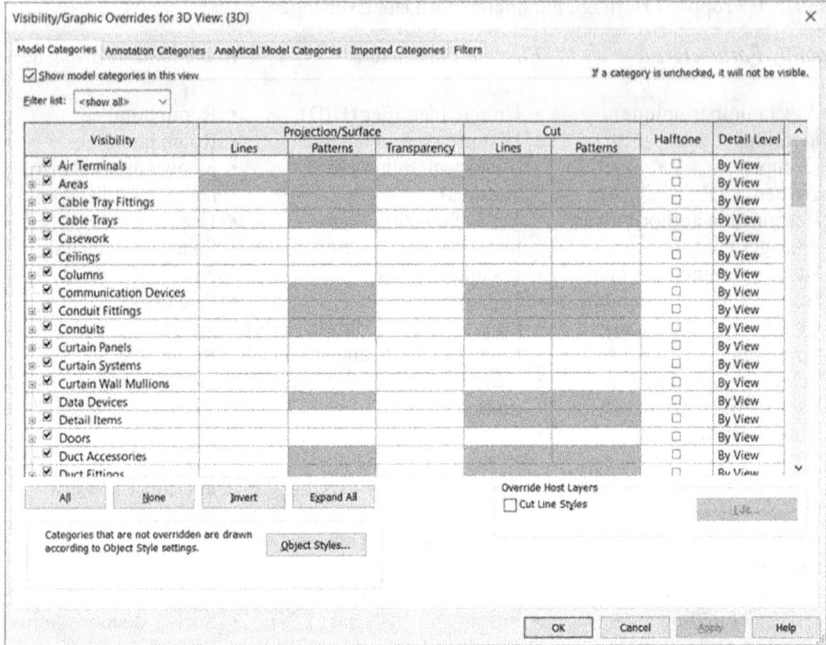

Figure 5.14 Unchecking Less Important Categories for VR Experience

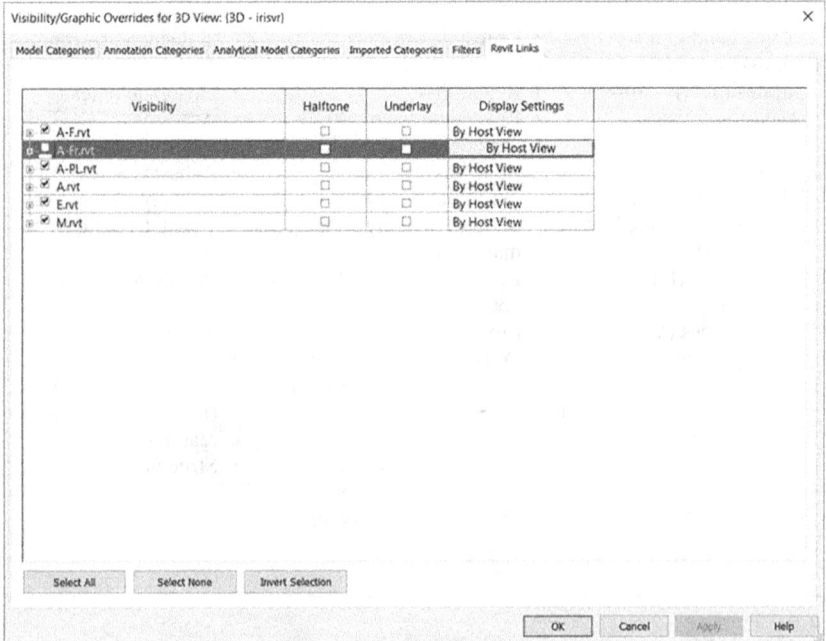

Figure 5.15 Removing Unnecessary Revit Links

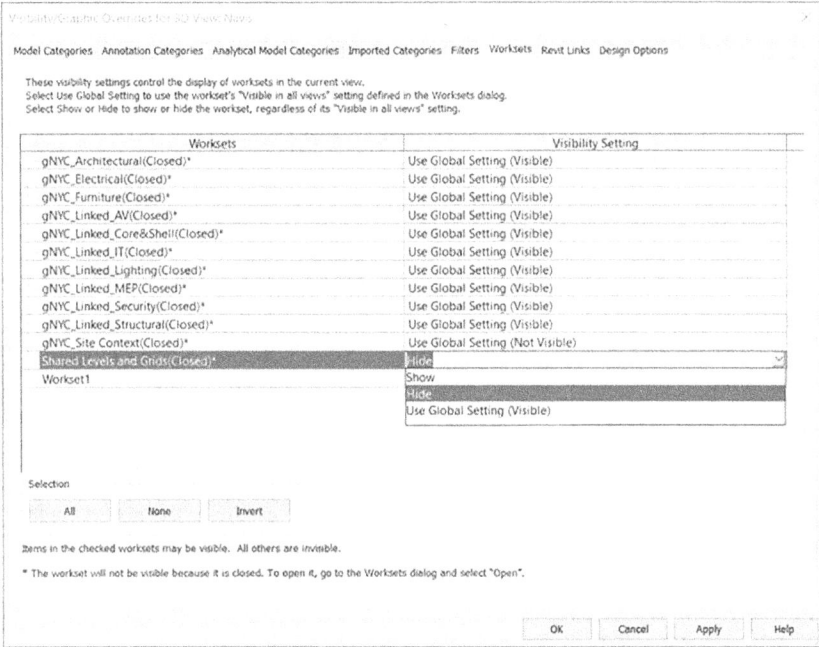

Figure 5.16 Hiding Unnecessary Revit Work Sets

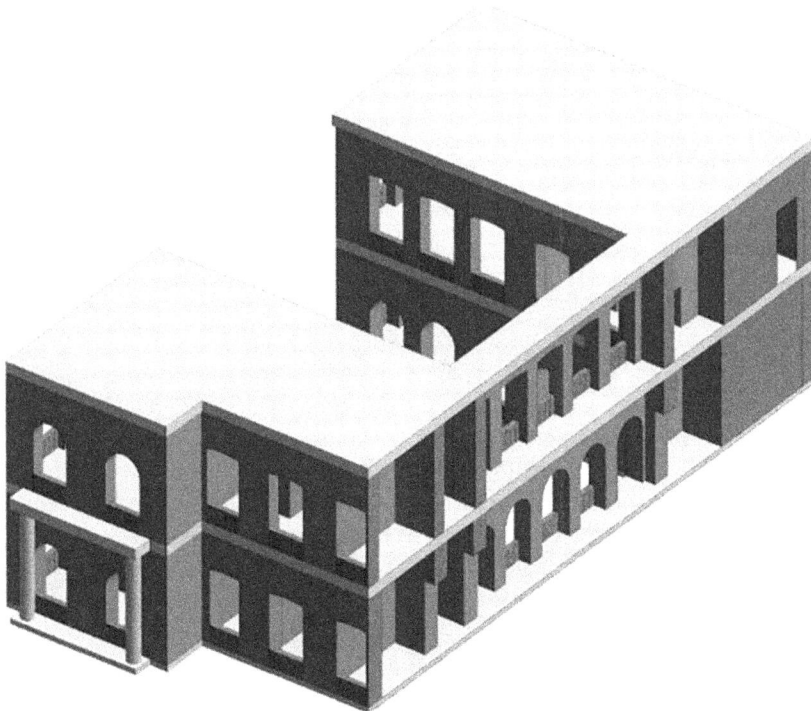

Figure 5.17 VR Model after Applying Section Box

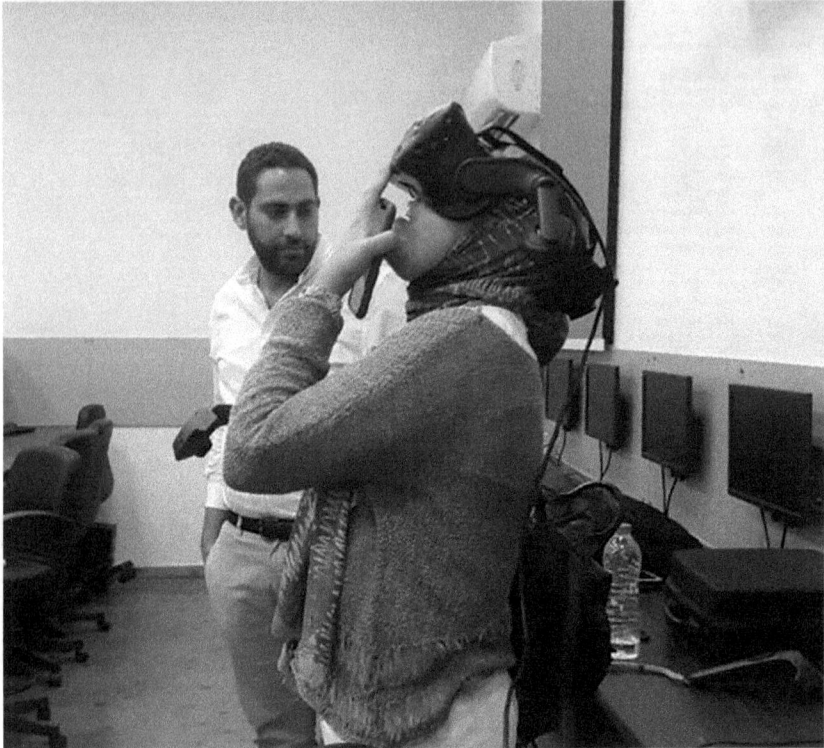

Figure 5.18 Using the HTC VIVE Pro System

The utilized VR system is the HTC VIVE pro system [12] for room-scale or standing setup. The basic system items are the headset for the immersive experience, two user interface controllers, and two base stations to monitor user movements (see Figure 5.18). Teleporting or fly-through movements may enable the user to navigate long distances in the VR model.

5.7   Summary

This chapter provided an overview of the proposed framework, including its main three phases: site exploration and scanning optimization, data acquisition and processing, and modeling the HBIM using the data. The first phase was divided into two stages; the first is site exploration, and the second is planning and optimizing the scanner's optimum locations. The second phase was divided into two steps. The first step is using a TLS scanner and a digital camera to scan and capture the texture of the heritage building according to the information gathered in the previous phase. The second step is processing the laser scanning data and the photography data. The third phase is divided into six steps which are modeling the main building skeleton, creating families for regular heritage elements, creating families for irregular heritage elements such as complex roofs, or domes, creating heritage parameters to the created models to form a full HAIM, and finally, visualizing the HBIM model using VR experience.

References

[1] M. Balzani, N. Santopuoli, A. Grieco, and N. Zaltron, "Laser scanner 3D survey in archaeological field : The forum of pompeii," in *International Conference on Remote Sensing Archaeology*, Beijing, pp. 169–175, 2004.

[2] K. Ikeuchi, T. Oishi, J. Takamatsu, R. Sagawa, A. Nakazawa, R. Kurazume, K. Nishino, M. Kamakura, and Y. Okamoto, "The great Buddha project: Digitally archiving, restoring, and analyzing cultural heritage objects," *Int. J. Comput. Vis.*, vol. 75, no. 1, pp. 189–208, 2007, doi: 10.1007/s11263-007-0039-y.

[3] M. Calin, G. Damian, T. Popescu, R. Manea, B. Erghelegiu, and T. Salagean, "3D modeling for digital preservation of romanian heritage monuments," *Agric. Agric. Sci. Procedia*, vol. 6, pp. 421–428, 2015, doi: 10.1016/j.aaspro.2015.08.111.

[4] A. Baik, "From point cloud to Jeddah Heritage BIM Nasif Historical House – case study," *Digit. Appl. Archaeol. Cult. Herit.*, vol. 4, pp. 1–18, 2017, doi: 10.1016/j.daach.2017.02.001.

[5] A. Frost, *Applied digital documentation in the historic environment*, 1st ed. Historic Environment Scotland, Edinburgh, 2018.

[6] S. Jayaram, H. I. Connacher, and K. W. Lyons, "Virtual assembly using virtual reality techniques," *CAD Comput. Aided Des.*, vol. 29, no. 8, pp. 575–584, Aug. 1997, doi: 10.1016/S0010-4485(96)00094-2.

[7] N. Dawood, "VR – Roadmap: A Vision for 2030 in the Built Environment," *J. Inf. Technol. Constr.*, vol. 14, pp. 489–506, 2009.

[8] B. Jiménez Fernández-Palacios, D. Morabito, and F. Remondino, "Access to complex reality-based 3D models using virtual reality solutions," *J. Cult. Herit.*, vol. 23, pp. 40–48, 2017, doi: 10.1016/j.culher.2016.09.003.

[9] E. Ramsey, "Virtual wolverhampton: Recreating the historic city in virtual reality," *Archnet-IJAR*, vol. 11, no. 3, pp. 42–57, 2017, doi: 10.26687/archnet-ijar.v11i3.1395.

[10] S. Deggim, T. P. Kersten, M. Lindstaedt, and N. Hinrichsen, "The return of the siegesburg – 3D-reconstruction of a disappeared and forgotten monument," in *International Archives of the Photogrammetry, Remote Sensing and Spatial Information Sciences – ISPRS Archives*, Nafplio, Greece, vol. 42, no. 2W3, pp. 209–215, 2017, doi: 10.5194/isprs-archives-XLII-2-W3-209-2017.

[11] "Reality Capture: Mapping and 3D Modeling Photogrammetry Software – Capturing Reality.com." [Online]. Available: https://www.capturingreality.com/. [Accessed: 22-Mar-2020].

[12] "VIVE™ | Discover Virtual Reality Beyond Imagination." [Online]. Available: https://www.vive.com/eu/. [Accessed: 22-Mar-2020].

6 Heritage Buildings Operational Performance

Maryam El-Maraghy

Researcher, Construction Engineering Technology Lab, Faculty of Engineering, Cairo University, Postal Code 12613, Giza, Egypt

Mahmoud Metawie

Assistant Professor, Structural Engineering Department, Faculty of Engineering, Cairo University, Postal Code 12613, Giza, Egypt.

Mohamed Marzouk

Professor of Construction Engineering and Management, Structural Engineering Department, Faculty of Engineering, Cairo University, Postal Code 12613, Giza, Egypt

6.1 General

The energy crisis and the increasing pollution worldwide have urged the need to rely on different sustainable energy resources and to reduce buildings' operational energy. Multiple building interventions work on reducing energy consumption and enhancing indoor environmental quality concerning indoor air quality, thermal comfort, and daylighting, as well as preserving the heritage value of the building. Recent BIM tools are incorporated to measure, assess, and evaluate enhanced building performance. Also, several assessment methods for analyzing and assessing the gathered data in order to enhance the buildings' operational performance. The multicriteria decision-making technique is one of the efficient techniques for prioritizing, ranking, and selecting the best sustainable interventions for predefined assessment criteria. A case study of a historic mosque in Cairo, Egypt, is presented to conduct a full energy analysis for assessing the current operational performance. Further, sustainable interventions are analyzed to enhance operational performance, utilizing a multicriteria decision-making model to choose the best intervention.

6.2 Buildings Operational Energy

Buildings' energy use is a major contributor to the worldwide rise in energy demand. As the world expands, more buildings are created, resulting in increased energy consumption attributable to structures. 20–25% of energy is consumed in developing countries and 30–40% in industrialized nations, increasing greenhouse gas emissions and resulting in 12% of water usage [1, 2]. According to IEA [3],

DOI: 10.1201/9781003357483-6

when indirect carbon emissions are considered, the building industry leaves a massive carbon footprint. Fossil fuel consumption in buildings accounts for almost 9% of global energy-related CO_2 emissions. Electricity and heating fuel used in buildings account for another 18%, while the production of building materials accounts for 10%. Therefore, as buildings account for about 37% of global CO_2 emissions during their entire lifecycles, whole life cycle emission controls are necessary [3]. That's why, with rising political, economic, and ecological threats like global warming, it's time to establish a cap on energy consumption [4].

As energy is consumed at every stage of a building's lifecycle, from construction to demolition, many already-existing buildings have experienced a decline in the efficiency of their building envelopes and mechanical and electrical systems, necessitating higher levels of energy use to maintain acceptable levels of indoor environmental quality [5]. Almost half of the energy used in a building in a developing country goes toward maintaining a comfortable indoor temperature, with the remaining 20% going toward internal loads such as lighting and appliances. An overwhelming majority (up to 77%) of a building's electricity usage comes from the heating, ventilation, and lighting systems. HVAC systems can consume up to 50% of a building's total energy in warmer weather, while lighting can consume up to 35% [6]. Numerous governments and international accords have identified the energy sector as the main target for enacting, especially in the direction of mounting environmental and climate change concerns. This has led to a rise in research into renewable energy as researchers look for more informed ways to guide the creation of a low-carbon economy and adopt sustainability practices [7].

Sustainable and energy-efficient buildings are the goal of many government initiatives and legislation. The environmental, social, and economic impacts of a building can collectively be used as metrics for its sustainability. Because of this, it is essential to use sustainability rating tools when assessing a building's sustainable performance [1]. Rating systems for green buildings were established by government agencies and nonprofit groups to minimize natural resource use and pollution. There are several green building rating systems in use around the world, including BREEAM (Building Research Establishment Assessment Method), LEED (Leadership in Energy and Environmental Design) in the United States, DGNB (Deutsche Gesellschaft für Nachhaltiges Bauen) in Germany, Green Star in New Zealand, CASBEE (Comprehensive Assessment System for Building Environmental Efficiency) in Japan, BEAM (Building Environmental Assessment Method) in Hong Kong, and BCA (Building and Construction Authority) in Singapore.. Notably, BREEAM is the first rating system for evaluating building performance based on defined target values for multiple categories [8, 9].

These rating systems and more initiative activities have been done to achieve more sustainable structures according to particular metrics and building types; nonetheless, there was a gap in taking heritage buildings and the variables impacting them into consideration in the assessment. Although several sustainability rating techniques are available, none are designed with HBs in consideration [1]. Heritage buildings are considered unique in their performance due to their old age and deteriorated materials, affecting performance, energy consumption, and life cycle costs. Hence, several factors should be assessed specifically for Heritage Buildings. Cultural, physical, digital, environmental, human, and social considerations and difficulties in social integration must be assessed during the restoration of historic structures [10]. Besides, the age and structural state of a heritage building, among other factors, may be important considerations in a sustainability analysis [1]. However, taking these assessment factors into account is not enough without identifying the factors that influence the operational performance of heritage buildings and the indoor environmental conditions.

The amount of energy required (or estimated) to keep a building operational is a key indicator of its efficiency. Environment, climate, orientation to the sun, on-site energy production, thermal comfort, and insulation are all factors in this calculation [11]. Heritage buildings' operational performance differs from other building types. These buildings are affected by their deteriorated building envelope, some weak structural elements, the presence of molds and other hazardous indoor environmental conditions, and many more factors that affect the operational performance concerning energy consumption, especially in the presence of HVAC system, the indoor environmental conditions, thermal comfort, and daylighting. These operational performance metrics are influenced by the building's age, status, and last renovation or maintenance.

Heritage buildings are increasingly seen as obsolete due to cracks that raise energy costs [1]. Heritage buildings typically have poorer energy efficiency and are far more difficult to access [10]. According to Al-Sakkaf et al. [11], heritage buildings are poorer energy efficient than contemporary ones, producing as much as 40% higher emissions for the energy consumed. Despite this, modern buildings utilize more materials, resulting in a net life cycle emission of only 8% less than those of historical buildings [11]. Therefore, it is critical to create well-articulated conceptions of heritage buildings' architectural and cultural qualities before beginning any renovations to increase their energy efficiency [4]. Building recycling and upcycling help lessen carbon emissions and other negative environmental impacts while simultaneously generating powerful urban regeneration projects [12].

In addition, several threats to irreplaceable artifacts are posed by the high indoor humidity levels and unpredictable temperatures in historic buildings. Hygrothermal interactions among the indoor environment, occupants, furnishings, and building exterior must be carefully considered. How a building element functions in unfavorable conditions is vital to the structure as a whole [13]. Microbial contaminants (e.g., mold and bacteria), gases (i.e., CO, CO_2, VOCs), and several other variables (such as temperature, humidity, and HVAC systems) all have an impact on the heritage building's indoor conditions [14].

In terms of lighting and daylighting, achieving a balance between artificial lighting and daylighting is one of the major challenges in illumination for heritage buildings [15]. The utilization of daylighting inside the space has its benefits regarding energy consumption reduction, internal heat gains, and achieving occupants' visual comfort. However, these are not the only concerns when it comes to heritage buildings, as other values should be taken into account, such as the heritage monuments and artifacts, also the indoor envelope materials [16]. This is because exceeding the daylighting limits could cause damage to the artifacts. Moreover, poor daylighting distribution, especially during hot weather conditions, could lead to more unwanted daylighting entering the space, causing visual discomfort [17].

6.3   Sustainable Interventions for Heritage Buildings

Existing buildings can be made more energy efficient, lowering energy costs, increasing investment returns, and slowing the rate at which energy consumption grows. According to Akande et al. [18], existing buildings can be upgraded to increase efficiency and operational performance at a lower cost than constructing from the beginning. Buildings that are designed to be sustainable save money on maintenance costs and use less energy since they generate their own power or employ renewable energy sources [2]. Each building is characterized by unique features that require a special retrofitting strategy to achieve the optimum building performance with the least possible energy consumption and better user comfort. Converting existing buildings to be sustainable ones is affected by different building parameters and the known level of information about the building to perform retrofitting actions for improving building energy use. Also, the available budget and circumstances for retrofitting actions all of that should be studied through the green retrofit methodology before implementation. As stated by Marzouk and Seleem [19], six main factors influence the energy performance of buildings, which are current climate, envelope characteristics, building activities, and energy systems, operation and maintenance of the building, occupants' activities and behavior, and indoor environmental quality. However, the most dominant factor that affects the performance of the building is the building envelope. Energy-saving design can improve the building envelope's physical features, resulting in lower operating costs and a reduction in carbon dioxide emissions [20].

The reason for retrofitting might be physical, functional, or both. The physical condition includes making the optimum use of available space and incorporating new technology, such as a solar roof and recyclable materials. Furthermore, the functional need covers the level of qualitative comfort in the indoor environment for users, such as heating, noise, visual, and energy usage [21]. An energy retrofit involves alterations to an existing building that changes the building's exterior appearance with the necessary approvals to raise efficiency. However, energy retrofits don't change its structural integrity, don't endanger the safety of neighboring constructions, don't introduce fire hazards, or harm the environment. It gathers numerous measures to improve the thermal envelope and reduce energy consumption for heating and cooling in existing facilities [22]. However, the difficulty occurs in deciding the points of weaknesses that need improvement; it needs to determine the factors that affect the building performance to define the possible retrofit strategies that could be implemented to enhance the performance of existing buildings.

The updating and adaptation of traditional buildings to meet present needs guarantee that they will continue to be used rather than being ignored and eventually demolished, making energy retrofits an increasingly popular preservation option [23]. In order to design strategies that can enhance heritage buildings' energy efficiency while maintaining their valued features, a comprehensive knowledge of their fundamental values is essential [4]. It is necessary to have a comprehensive understanding of how these structures function after the completion of the renovation and reuse projects to face the pervasive assumptions that discourage energy efficiency investments in the historic building sector. Understanding how even little changes, like switching to energy-efficient light bulbs or improving service efficiency or building operation (i.e., facility management and the building usage pattern of staff and other users), can help overcome these problems is also crucial [18].

Energy efficiency, comfort, and the building's original aesthetic and historical significance should all be prioritized during the retrofitting process for historical buildings [24]. Retrofitting heritage buildings can be challenging for two main reasons: physical features, number one. These structures often employ nonmechanical, passive techniques of indoor climate control, such as thermal mass and natural ventilation through openings. They may have complex and irregular geometries and envelope construction without insulation or vapor barriers. The second premise is that of conservatism. Conservation principles and practices regulate how historic buildings should be handled; this includes safeguarding the structure's original materials and design as much as possible [23]. Various studies applied and proposed some interventions to enhance the buildings' thermal comfort, daylighting, and energy performance and consider cultural value. Akande et al. [18] listed some sustainable interventions to be used in reusing and refurbishing historic churches, such as improving and replacing the heating system, utilizing a renewable source of energy, regulating energy consumption and user behavior, adding underfloor heating systems, upgrading the building's energy efficiency by getting rid of old, inefficient machinery, and applying insulation layers for walls and windows. They added that these interventions could be applied to other historical buildings taking into consideration the building's current condition. In order to achieve indoor thermal comfort as well as preserve the heritage value, Ibrahim et al. [25] proposed passive interventions for a heritage residential building in Cairo; these interventions included mixed mode ventilation (using natural ventilation via windows and cooling system), solar control (using Low-E double glazing and cooling roof through applying white paint), thermal control (using different insulation layers for the roof and walls), and hybrid scenarios (integration between more than one intervention) [25]. Also, Marzouk et al. [26] studied the effect of changing the skylight configuration of a heritage building in Cairo on the energy and daylighting performance of the building. The results revealed that the cooling load could be reduced by 6.5%, and an average of 133% could enhance useful daylighting illuminance.

These interventions should be assessed after that according to well-defined metrics to indicate their efficiency and determine their impact on the operational performance of the building. Accordingly, Webb [23] grouped the performance metrics to assess the impact of the sustainable interventions on the buildings' performance into four main categories with their associated criteria as follows:

1 **Global environment**: it includes energy consumption (annual energy consumption, annual CO_2 emissions), energy production, climate change vulnerability (exposure to hazardous conditions, adaptive capacity), and embodied energy.
2 **Building fabric**: it includes conservation (i.e., visual impact on heritage value, material impact on heritage value, compatibility) and hygrothermal behavior (i.e., thermal transmittance, thermal mass, moisture).
3 **Indoor environment**: it is related to occupants (i.e., indoor environmental quality: thermal comfort, indoor air quality, lighting) and collections (i.e., agents of deterioration: pollutants, incorrect temperature, incorrect relative humidity).
4 **Economics**: it counts for the economic costs of these interventions and their impact (i.e., capital costs, operational energy costs, maintenance, and replacement costs).

However, to be able to use these metrics to assess the current condition of the heritage buildings to determine the areas of weaknesses that need improvement as well as evaluate the impact of the sustainable interventions, the building data (i.e., geometrical, physical, and indoor conditions) should be acquired and assessed using the suitable performance assessment method to enhance its operational performance.

6.4   BIM and Data Acquisition Tools

Architectural assessments of historic buildings can be easily carried out using nondestructive imaging technologies and long-range instruments like 3D terrestrial laser scanning and photogrammetry. BIM tools and 3D architectural models could be used for the calculation of alternative designs, estimation of the required costs, quantification of materials, documentation of buildings through creating as-built models to be used for retrofitting existing buildings or creating new ones, managing data, and others [27]. 3D scanning has strongly influenced precision and offered an almost limitless set of possible uses. Due to its ability to rapidly collect massive volumes of data quickly, this would be a much superior method than any previous measurement system in use [28]. Moreover, it has several advantages compared to traditional surveying methods (e.g., data acquisition speed, large data volume, high precision, ability to work in various situations, and a wide range of applications). This untouched measurement system obtains 3D coordinates, reflection intensity, and the like on object surfaces [29]. A terrestrial laser scanner can be used in construction and architecture industries for many purposes. Documentation such as 3D models, sectional elevation views, and multimedia visualizations can all be created using this technology. With terrestrial laser scanning (TLS), it is feasible to produce detailed documentation for conservation and restoration projects aimed at preserving or even reconstructing damaged architectural features in the preservation of historic sites [30].

There has been a large number of research, both theoretical and practical, that shows that 3D laser scanning is an efficient and accurate tool for acquiring building data for further assessments [27, 30–32]. Previtali et al. [33] developed a 3D BIM model of a building façade built from laser scanning point clouds with the inclusion of infrared thermographic analysis to examine the energy efficiency of buildings and discover thermal anomalies. As a part of digital documentation for heritage buildings, Jo and Hong [32] used terrestrial laser scanning and unmanned aerial vehicle (UAV) photogrammetry and digital recording of the structure's interior to build a three-dimensional model of Magoksa Temple in the Republic of Korea. 3D Laser scanning and UAV photogrammetry were used to acquire perpendicular geometry of the buildings and places. In contrast, UAV photogrammetry provided greater planar data gathering rates in upper zones, such as the top of a building. Photogrammetry and laser scanning were compared using ground control points, and it has been proven that laser scanning is more accurate [32]. 3D Laser scanning and UAV photogrammetry were shown to operate well in this study to digitally document and analyze cultural heritage sites. Rocha et al. [27] used digital photogrammetry and 3D terrestrial laser scanning to conduct an architectural examination of the Instituto Superior de Agronomia in Lisbon in order to repair the building. The purpose of the case study was to demonstrate how surveys can be conducted and data collection methods are modeled. In a study by Gardzińska [30], the old Jaroslaw market in Poland was used to demonstrate how terrestrial laser scanning (TLS) technology can be used to create 3D models of "greenfield" structures such as walls and buildings. The study illustrated how this technology could be used to document conservation and restoration activities to maintain monuments in their current state or even restore damaged architectural components [30].

The use of new technologies, such as infrared thermography, has a vital role in determining the building envelope characteristics and the buildings' defects, affecting the heritage buildings' operational performance. Where the surface temperature of an object can be measured with infrared thermography (IRT), a form of noncontact and nondestructive testing [34]. According to Paoletti et al. [35], IRT flexibility allows it to be utilized in various contexts, providing additional insight by making previously hidden information visible. Its usefulness has been demonstrated in various contexts, including predicting structural failures, analyzing humidity issues, and estimating urban heat loss. As a diagnostic tool, it examines the thermal behavior of building components to spot anomalies and decaying areas [34]. It is also used to evaluate the thermal performance of a building's windows, walls, and other components [36]. Using infrared thermography, Ghosh et al. [34] assessed the current condition of a heritage building by investigating and locating envelope damage, such as cracks and moisture damage. Also, to determine the envelope performance in a heritage building, Wang and Cho [37] combined two data acquisition technologies: laser scanning and thermal imaging. They created a hybrid 3D LIDAR system that incorporates an IR camera to measure a building's surface temperature; this allows the temperature data to be instantly integrated with the specified scan points during data acquisition, and for each point in the cloud to be defined by its x, y, and z coordinates and temperature data. The data are entered into a building performance simulation model to assess energy efficiency [37].

6.5   Operational Performance Assessment Methods

Inspecting a historic structure is essential for determining the quality of the building's envelope and the indoor environment. The envelope characteristics are often unknown, making tests and field data indispensable for determining the current behavior of a building. It is crucial to evaluate the thermal resistance of heritage buildings' exterior and interior envelopes to model the complete structure and examine different retrofit alternatives [36]. The effectiveness of a retrofit to a heritage place can be estimated or measured using several different approaches. Buildings' energy performance can be evaluated by calculating the amount of energy needed, the technologies employed, and the facility's use, all of which can be derived from a thorough characterization and quantification of their energy consumption sources. Thus, numerous nations have created and implemented energy performance evaluation methodologies to aid in selecting choices that will increase building energy efficiency [5]. Moreover, as mentioned by Webb [23], there are different methods used in previous studies to analyze and assess the impact of the sustainable interventions on the overall performance, which include *microclimate analysis* to assess the indoor environmental conditions (i.e., temperature, relative humidity) and their impact on the historical artifacts and occupants' thermal comfort, *building envelope measurement* to determine the thermal characteristics of the building envelope (i.e., U-value) either pre-retrofit or post-retrofit, *energy simulation model* to simulate the buildings' current conditions and establish a baseline to be used as a reference for assessing the impact of the retrofit measures on the operational performance, *Hygrothermal simulation* to assess the factors that affects the indoor moisture that results from conditioning equipment, ventilation, infiltration, and vapor transfer*, defects detection* to monitor and assess the damage and deterioration of the buildings' materials as a result of biological, chemical, or mechanical material degradation.

Also, to assess the building's structural performance, Structural Health Monitoring (SHM) to assess the structural performance of the building as well as to ensure the structure's safety and preserve the architectural significance of the building [38]. It is conducted by gathering field data to assess the structural performance. If the damage is discovered in time, the building can be reinforced or retrofitted such that it does not collapse [38]. Besides, global environment indices are used to assess and predict the buildings' climatic conditions. Air temperature, relative humidity, and humidity mixing ratio inside buildings can be forecast for the future using estimations of climate change and models of the buildings [13]. For Example, Huijbregts et al. [39] evaluated the impact of climate change on the typical indoor environment of historical museums in the Netherlands and Belgium, as well as the potential damage to museum artwork. To determine the potential risks that climate change poses to museums and their collections, the authors analyzed the indoor environment and the museum's holdings using the (ASHRAE) guidelines for museum collections and Marten's specialized climate risk assessment model for museum artifacts.

6.6   Buildings Performance Evaluation Using MCDM

Decision-makers use multicriteria decision-making (MCDM) techniques to evaluate different alternatives for a certain problem using predefined criteria given certain weights to obtain the best options. As mentioned by Karmaker et al. [40], multicriteria decision-making models are widely acknowledged as among the most effective methods when it comes to ranking options and making a final decision. Solving decision-making problems using these steps is achieved using different MCDM techniques that were established years ago and used in solving complex assessment problems. MCDM techniques were widely used in solving problems regarding sustainability and building fields such as buildings' performance, technology selection, and energy planning and policies [7, 41, 42]. These techniques are Analytical Hierarchy Process (AHP), Analytical Network Process (ANP), Technique for Order Preference by Similarity to Ideal Solution (TOPSIS), Preference Ranking Organization METHod for Enrichment of Evaluations (PROMETHEE), Elimination and Choice Transcribing Reality (ELECTRE), and the Fuzzy set theory. The AHP technique is one of impact analysis's most commonly used techniques. Together with TOPSIS, PROMETHEE, and Fuzzy Set, they are the most commonly used methodologies in research, especially technology evaluation [7]. Each technique has different characteristics in solving the decision-making problem [43]. AHP divides the decision-making problem into a hierarchy of criteria which are weighted according to the decision-makers' judgments and preferences; the TOPSIS, ELECTRE, and PROMETHEE techniques are applicable to be used in decision problems that include few criteria with a large number of alternatives [24, 42–45].

Using the MCDM techniques, the decision-maker can evaluate the efficiency of various retrofit measures in terms of sustainability. Stanojević et al. [22] used the fuzzy AHP to prioritize energy performance indicators for retrofitting heritage buildings in Serbia. Roberti et al. [24] evaluated different retrofit alternatives for a historic building in Italy are evaluated and selected using AHP and multiobjective optimization based on achieving three main objectives, which are decreasing the energy demand, increasing thermal comfort, and conserving the heritage aspect. The AHP model was also used by D'Alpaos and Bragolusi [46] to prioritize energy-efficient retrofit strategies in public housing. The model results showed that priority should be given to comfort benefits and thermal insulation, lowering the extra capital costs and delivery costs throughout the building's lifetime. Si et al. [43] used the AHP technique for assessing and selecting green retrofit technology for a historic university building in London based on three criteria: annual energy savings, payback period, and investment costs.

Using the PROMETHEE technique, Dirutigliano et al. [47] ranked five buildings' retrofit alternatives from two different perspectives: the public and private sector perspectives. At the building scale, the owner is the decision-maker who selects the appropriate retrofit strategy for his building. In contrast, at the district scale, the decision-maker is the municipality, which ranks the proposed retrofit alternatives that provide 20% energy savings. The ELECTRE technique was used by Dall'O' et al. [44] to evaluate and rank five retrofit strategies with different energy retrofit actions to be applied to existing buildings in Melzo, Italy. According to Tan et al. [45], the AHP and TOPSIS techniques are the most widely used in MCDM, where the AHP method is used for calculating the weights of the decision criteria while the TOPSIS is used for ranking the alternatives.

6.7   Case Study

Being one of the heritage buildings in Egypt, Imam El-Hussein Mosque was adopted as a case study to be assessed and evaluated following its operational performance. The mosque's operational performance was assessed based on a proposed framework that involves four main stages. The first stage, the data acquisition, involves collecting the mosque's data using 3D laser scanning and a site investigation to define the different building's characteristics, such as the building envelope, and installed systems. After that, the second stage, building simulation model development, involves creating a BIM model for the mosque using the processed point clouds and then performing a detailed analysis of the mosque's operational performance using energy simulation software to define the areas of weaknesses in the mosque. Then, in the third stage, some sustainable interventions are applied to assess their impact on the operational performance of the mosque. And finally, to determine the most sustainable intervention following the mosque's performance, two decision-making techniques were used AHP and TOPSIS. The AHP was used to determine the weights of the different assessment criteria. After that, the TOPSIS was used for ranking the different interventions to choose the best among them. The details of these stages' implementation in the mosque are described in the following subsections.

6.7.1   Case Description

Imam Al-Hussein Mosque is located near the Al-Azhar Mosque in Cairo, Egypt. It is named after Imam Al-Hussein ibn Aliand is one of Egypt's holiest Islamic shrines. The mosque was erected during the Fatimid reign in 1154 AD and later reconstructed in 1874. It is presently included in Historic Cairo, designated a UNESCO World Heritage Site in 1979 [48]. Figure 6.1 shows the exterior Imam El-Hussein Mosque.

6.7.2   Mosque Data Acquisition

Since the Imam El-Hussein Mosque was built hundreds of years ago, collecting its geometrical data (such as its dimensions, partitions, and all other relevant and important architectural data) using the conventional methods of documentation – surveying, measuring, and drawing – can be time-consuming and mistaken. Hence, the mosque's geometrical data was acquired using 3D laser scanning. The process

Figure 6.1 Imam El-Hussein Mosque

involved site investigations for making the preliminary measurements and creating a detailed site plan to select the best scanning locations for accurately scanning the entire building. A total of 236 laser scans were taken of the entire mosque using a Z+F IMAGER® 5010C laser scanner. Point clouds were produced from these scans, which are then processed (to be filtered from any unwanted data) to be imported into Autodesk Revit for creating the BIM model. Figure 6.2 shows the point clouds of the mosque after being processed. Mosque's performance data was gathered through the energy audit. The energy audit was conducted through surveys, interviews with the mosque's operators, and site investigation. The gathered data involves the operation schedule, occupancy pattern, installed lighting system (type, number), the utilized ventilation method, current building constraints, and the building envelope characteristics.

6.7.3   *Mosque Simulation Model Development*

A BIM model for the whole mosque (see Figure 6.3) was constructed in Autodesk Revit software, using the processed point clouds to develop the mosque's main structural and architectural elements. The BIM model after that was used for detailed operational analysis using (Integrated Environmental Solutions-Virtual

Figure 6.2 Processed Point Clouds for the Whole Mosque

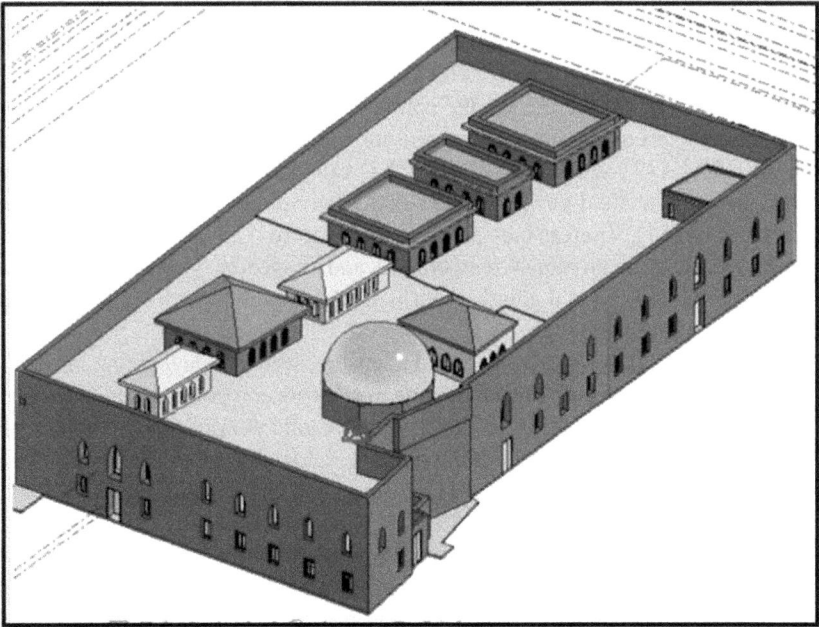

Figure 6.3 Generated BIM Model for the Mosque using Autodesk Revit Software

Environment) software [49]. The data gathered from the energy audit are defined within the software for conducting a detailed energy analysis to define the current mosque's performance and the areas that need enhancement. The defined inputs include (see Table 6.1) (1) defining the current mosque location to account for climatic conditions during the analysis, (2) identifying the prayer area that will be analyzed, and (3) adding the factors that affect the mosque's operational performance and that are collected during the energy audit, which are (a) the building envelope characteristics, (b) the installed lighting and electrical systems data, and (c) the occupancy pattern and operational profile. More details about the case study's Base Case (BC) can be found elsewhere [50].

The Base Case (BC) is then analyzed with respect to four main metrics, which are:

1 *Energy Performance* which is assessed through the annual energy consumption (MWh) and the associated carbon dioxide emissions ($KgCo_2$).
2 *Building heat gains* which are divided into internal heat gains and external heat gains, where the internal heat gains are the loads generated within the building, such as the number of people, ventilation rate, lighting, equipment, and operation schedule [6], while External heat gains in space are caused by sun radiations, which enter through apertures such as skylights and windows or indirectly through building envelope components such as roof and walls.
3 *Thermal comfort is one of the dominant factors affecting* a building's occupants [49]. It is assessed inside the prayer areas with respect to three metrics: the Predicted Mean Vote (PMV), the Percentage of People Dissatisfied (PPD), and the Operative Temperature. The PMV is one of the indices that are widely used for evaluating the occupants' comfort was proposed by Fanger [50] and is measured through a seven-point sensation scale (−3 to +3) that is proposed by ASHRAE [49], while the PPD, which is calculated based on the PMV, indicates how dissatisfied people in the existing environment are with the current temperature [51], whereas the operative temperature is calculated by combining air temperature, mean radiant temperature, and airspeed to generate a simplified metric of human thermal comfort that is used for analyzing the comfort of the building's occupants [49].
4 *Daylighting performance* is addressed to ensure sufficient daylighting inside the room and visual comfort. The daylighting metrics are composed of static daylighting metrics, such as illuminance level, and dynamic metrics that are commonly used for evaluation, such as DA (Daylighting Autonomy), sDA (Spatial Daylight Autonomy), ASE (Annual Sunlight Exposure), and UDI (Useful Daylight Illuminance) [52]. The metrics used for the Base Case evaluation are illuminance level (lux) as a static metric to determine the average illuminance value in the three prayer areas. In contrast, the dynamic metrics are the sDA which represents the percentage of occupied time when daylight meets the specified illuminance at a place in a space [53], and the ASE to assess the amount of visual discomfort inside the space. According to IESNA [54], it is the proportion of an area utilized for analysis that is exposed to direct sunlight above the permitted illuminance level during the specified number of occupied hours per year.

Table 6.1 Base Case Parameters

Parameter	Description	
Building envelope characteristics	*Windows*:	Single glazed windows, Thickness = 3 mm, U-value = 5.906 W/m^2 K)
	Roof:	Thickness = 360 mm, U-value =1.0653 W/m^2 K.
	Walls:	Thickness= 800 mm, U-value= 0.6131 W/m^2 K
Ventilation method	Naturally ventilated through windows and installed ceiling and wall-mounted fans	
Lighting system	Large chandeliers and lighting units with CFL lamps (23 W and 13 W).	
Electrical systems	Speakers (300 W), ceiling fans (50 W), wall-mounted fans, and vacuum cleaners	
Operational profile	Daily from 3:00 to 4:30 am, 8:30 am to 10:00 pm (summer)/ 9 pm (winter)/11:30 pm (Ramadan)	
Occupancy pattern	25% to 50% during daily prayers, 100% during Friday prayers and Ramadan (Ishaa Prayer)	

The existing state of the mosque and its operational performance were assessed using the stated metrics. The analysis revealed that there are areas of strength and areas of weaknesses in the current operational performance. Concerning *energy performance*, the annual energy usage is 331.7 MWh. The lighting system is considered the mosque's largest energy consumer. This occurs due to the huge chandeliers and lighting equipment installed in mosques and their constant operation throughout the day. The lighting system consumes 274.36 MWh (88%) of the total energy consumed, whereas the other systems consume 37.87 MWh (12%). Moreover, the annual energy use results in carbon dioxide emissions of a value of 172,166 KgCo$_2$.

Concerning the *building heat gains*, the internal heat gain average for the prayer area is 1,108 MWh. People account for 74% of total internal heat gains in a mosque accommodating around 3,000 persons. The lighting system is the second largest contributor to internal heat increase, accounting to 23% of the total. Other appliances contribute approximately 3% of total internal heat gains. Whereas, regarding the external heat gains, the prayer area has an average external heat gain of 345 MWh. Reducing solar heat gain is a primary priority in warm and hot climates due to its significant impact on interior temperature, which impacts cooling loads and thermal comfort.

As for the *thermal comfort*, the PMV values were between −1 and +2.96 throughout the year, with an average value of +1.34 during the summer months (May–October), which means that they are more than the acceptable limits that should be from +1 to −1 as stated by Atmaca and Gedik [51]. During the summer months, the average percentage of discomfort hours (PMV > +1) is 68.8% of total occupied hours. In addition, the PPD values in the prayer areas reached 98%, with an average value of 40%. Also, the percentage of discomfort hours in which the PPD > 20% is 76.9%. Furthermore, the operational temperature exceeds 40°C during summer, averaging 28.5°C. The percentage of discomfort hours when the operational temperature surpasses 26°C is 72%. The thermal comfort results in PMV, PPD, and operative temperature refer to uncomfortably hot temperatures in the prayer areas, particularly during summer.

Regarding *daylighting performance*, in general, 100 Lux is the lowest allowed amount regarding the illuminance level for mosques [55, 56]. At a working distance of 0.4 m, the average illuminance during the four extreme daylight times is used to model the lighting in the three prayer areas. The prayer area has an average illumination of 395.1 lux. Figure 6.4 shows the average illuminance distribution inside the prayer area. Moreover, concerning the sDA, since the range of allowable

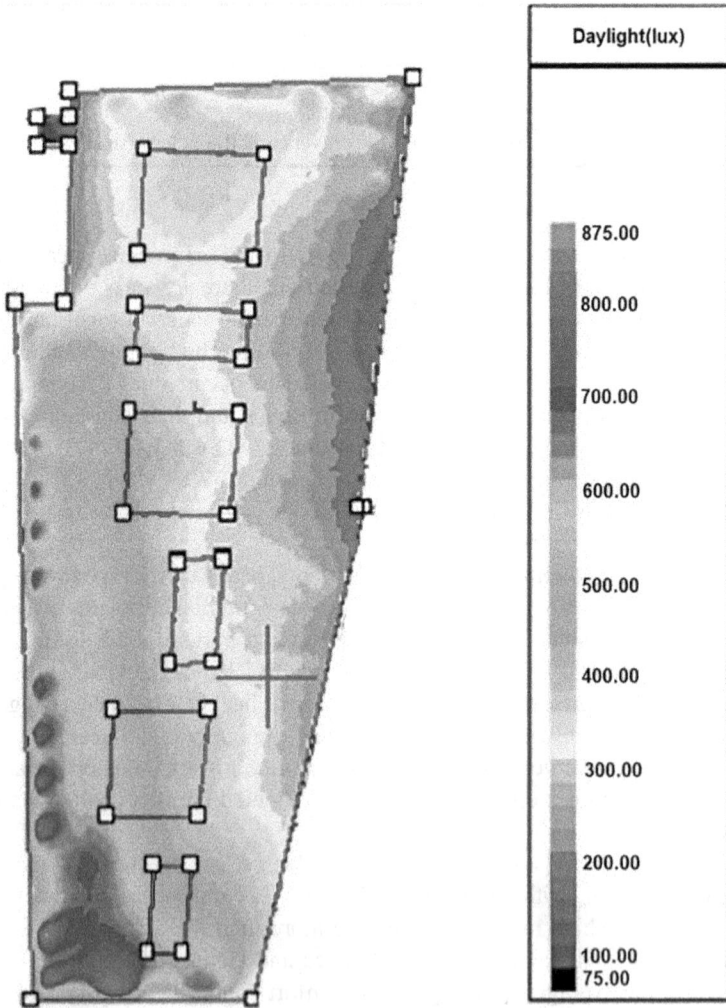

Daylight(lux)
875.00
800.00
700.00
600.00
500.00
400.00
300.00
200.00
100.00
75.00

Figure 6.4 Illuminance Distribution Inside the Prayer Area

illumination in mosques is 100–300 Lux, the mosque is evaluated to guarantee that the 300-lux threshold was met during 50% of the occupied hours. More than 50% of the occupied time was under conditions when the sDA was greater than 300 lux. The sDA is 85.39% on average. Figure 6.5 shows the sDA distribution inside the prayer area. On the other hand, the average ASE value for the three prayer areas has exceeded the acceptable limit, where ASE is defined as the proportion of area that surpasses 1,000 lux at a given position for more than 250 occupied hours by IESNA. It should be less than 10% to be acceptable and avoid glare. But the average ASE value is 64.7%. Figure 6.6 shows the ASE distribution inside the prayer area.

6.7.4   *Sustainable Interventions Impact*

Based on the results of the Base Case evaluation and the observed weaknesses areas that need to be addressed to improve the mosque's operational performance, the proposed sustainable interventions are divided into the following four groups: (1) changing the building envelope characteristics (using double glazing, triple glazing, and roof insulation layer), (2) changing the lighting system (using LED lamps and dimmers), (3) changing the ventilation method (using cooling system and opening skylights windows), and (4) changing the operation schedule to be

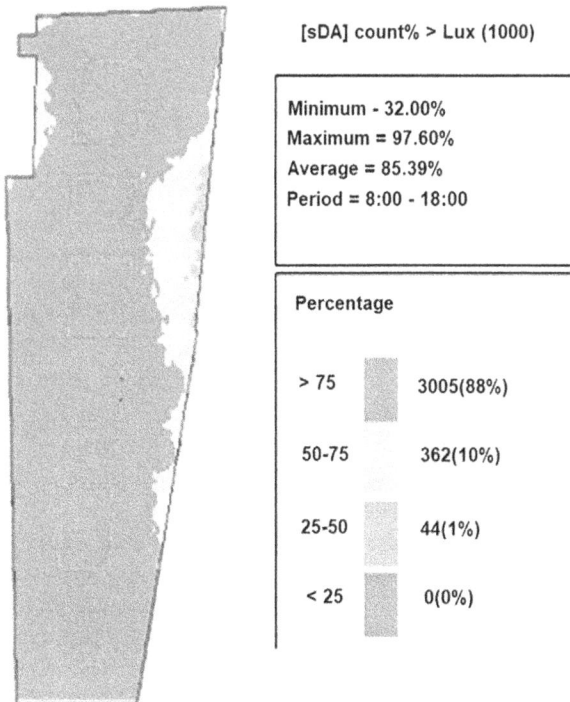

[sDA] count% > Lux (1000)

Minimum - 32.00%
Maximum = 97.60%
Average = 85.39%
Period = 8:00 - 18:00

Percentage

> 75   3005(88%)

50-75   362(10%)

25-50   44(1%)

< 25   0(0%)

Figure 6.5 sDA Distribution Inside the Prayer Area

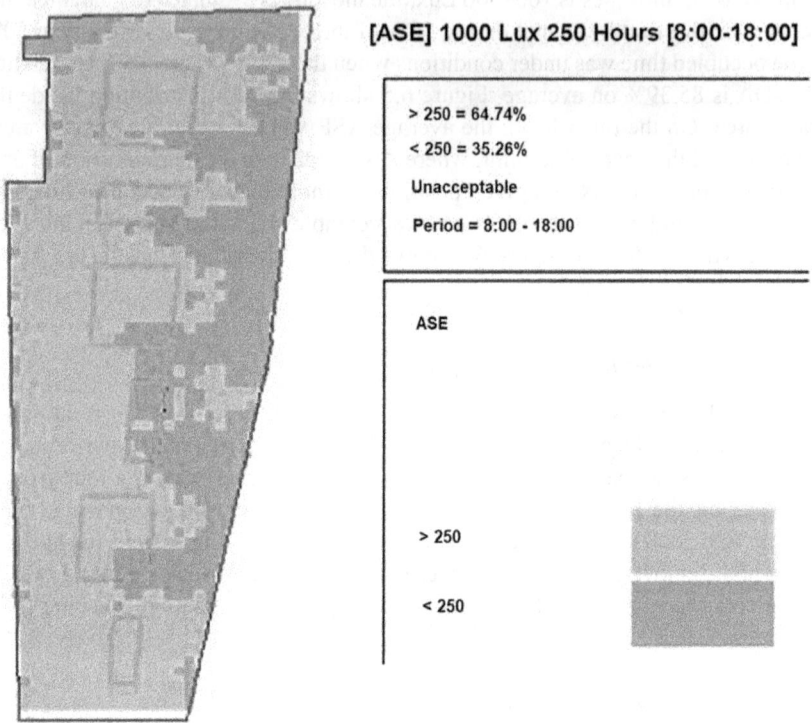

Figure 6.6 ASE Distribution Inside the Prayer Area

operated one hour with each prayer and two hours for Friday and Ramadan. By combining and/or separating these categories, a wide variety of intervention scenarios can be generated. The proposed seven retrofit scenarios are listed in Table 6.2 due to combining various retrofit methods.

Results for each retrofit scenario concerning the stated metrics are compiled and shown in Figure 6.7a–j, which compares the scenarios to the Base Case in terms of energy performance, thermal comfort, daylighting performance, and building heat gains.

Some of these scenarios had a favorable influence on certain metrics while negatively impacting others; the outcomes of the retrofit scenarios with respect to the defined metrics are discussed below.

Concerning the *energy performance,* as shown in Figure 6.7a, the highest energy consumption value was obtained in Scenario 3, where it increased from 331.7 MWh as in the Base Case to 796.76 MWh (+140%) because of adding the cooling system. However, by adding other retrofit measures, such as using LED lamps with dimmers and changing the operation profile with the cooling system, double glazing, and roof insulation layer, as in Scenario 7, the annual energy consumption value decreased to 257 MWh. Replacing the current lamps with LED lamps, as

Table 6.2 The Studied Retrofit Scenarios

Scenario ID	Description
S1	Replacing single glazed windows with double glazing windows (glazing characteristics: 6-14-6 mm double glazing window – argon filled with a reflective coating, U-value = 1.4867 W/m² K)
S2	Replacing the current CFL lamps with LED lamps (10 W)
S3	Adding a cooling system (Cooling System: multi-splits AC units)
S4	Changing the operation profile to be operated one hour with each prayer and two hours with Friday and Ramadan
S5	Replacing single-glazed windows with triple glazed windows ((6-14-6-14-6 mm) triple glazing window – argon filled with a reflective coating, U-value = 1.0503 W/m² K), changing the operation profile, and using a roof insulation layer (thickness = 100 mm, conductivity = 0.028 W/(m².K), the new roof U-value = 0.2068 W/m² K)
S6	Using LED lamps with dimmers (to decrease the lighting intensity during the presence of daylighting) and using double-glazed windows.
S7	Using LED lamps with dimmers, using double-glazed windows, adding a roof insulation layer, using a cooling system, and changing the operation profile

Note: The software calculates all U-values

in Scenario 2, decreased the energy consumption from 331.7 MWh to 198 MWh. Moreover, using LED lamps with dimmers, as in Scenario 6, yielded a reduction in energy consumption by 73%, the lowest result among the simulated retrofit scenarios. The percentage decrease in carbon dioxide emissions is proportional to the percentage change in the building's energy consumption. Emissions are highest in Scenario 3, at +140% above the Base Case, and lowest in Scenario 6, at −73% below the Base Case. Figure 6.7b shows the values of the carbon dioxide emissions associated with the amount of energy consumed for the different retrofit scenarios.

For the *thermal comfort,* as shown in Figures 6.7c–e, incorporating the cooling system in Scenario 3 without making other modifications to the Base Case considerably increased the thermal comfort of the prayer areas' occupants during the summer months. For PMV > 1, the percentage of discomfort hours dropped from 68.8% to 10%, and the average PMV decreased from 1.34 to 0.68. Moreover, for PPD, the percentage of discomfort hours (PPD > 20%) dropped from 76.9% to 14.4%, and the average PPD decreased from 45.55 to 17.35. Also, the percentage of discomfort hours when the operative temperature is >26 dropped from 72% to 19%, and the average operative temperature decreased from 28.47 to 24. When adding other retrofit interventions with the cooling system, such as LED lamps with dimmers, double glazing, roof insulation layers, and changing the operation profile, it yielded the lowest values, with the percentage of discomfort hours being 6.4%, 7.3%, and 7.9% for PMV > 1, PPD > 20, and OT > 26, respectively.

Concerning the *building heat gains,* with no alterations to the operation or electrical systems, the greatest internal heat gains value is 1,180 MWh, which was reached in the Base case (BC) and other scenarios like Scenario 1 and Scenario

3 (see Figure 6.7f). On the other hand, changing the mosque's operation profile resulted in a 56% reduction in internal heat gains from 1,180 MWh to 518.6 MWh across Scenarios 4 and 5, with no other changes to the mosque's electrical systems. And the lowest internal heat gains value was obtained in Scenario 7, which is 464 MWh when the LED lamps with dimmers were replaced and the operation profile changed. Regarding the external heat gains, as shown in Figure 6.7g, Base Case, Scenarios 2–4, where no alterations were made to the building envelope, yielded the maximum value of 345 MWh. On the other hand, when double-glazed windows were used, as in Scenarios 1 and 6, the heat gains value decreased from 354 to 260.8 MWh. Moreover, Scenario 5, which considers using triple-glazed windows and adding a roof insulation layer, has the lowest value, at 208 MWh (a 39.7% reduction).

Concerning the *daylighting performance*, using double-glazed windows with reflective coating as in Scenarios 1, 6, and 7, the average illuminance level decreased from 395.1 to 263.8 Lux. Moreover, on using triple-glazed windows with a reflective coating, as in scenario 5, the average illuminance dropped from 395.7 to 220.7 Lux, slightly below the desired preferred, acceptable range (250 Lux), as shown in Figure 6.7h. The daylighting values in Scenario 5 are the lowest of all the alternative scenarios. Figure 6.7i shows the average sDA values for all of the scenarios. The average sDA value dropped from 85 to 76.76% in Scenarios 1, 6, and 7. However, it is still in the preferred, acceptable range (more than 75%). Scenario 5 yielded the lowest sDA value of 58.4%; however, it is still acceptable (more than 50%). Besides, the lowest ASE was obtained in Scenario 5 with a value of 26.9%, as shown in Figure 6.7j.

As shown in the results, the impact of the retrofit scenarios on the operational performance of the mosque varies for the different metrics; some yielded a positive impact concerning some metrics and a negative concerning others and vice versa. Therefore, deciding which retrofit scenario is the best based on its impact on operational performance with respect to the defined assessment metrics is challenging.

6.7.5 Assessment of the Sustainable Interventions

The best-retrofit scenario can be selected according to the defined assessment criteria using MCDM. The AHP and TOPSIS techniques are used to evaluate and rank the various scenarios. What follows is a description of the three stages of the MCDM process:

6.7.5.1 Stage 1: Identification of the Evaluation Criteria

Environmental impact, thermal comfort, daylighting performance, building heat gains, and life cycle cost are the proposed main criteria for evaluating the defined scenarios. Each of the defined criteria in Table 6.3 has multiple attributes. Results from the energy simulation model are used to determine values for the first four criteria. Whereas the values of the lifetime cost and lifetime energy savings are computed based on the expense of implementing the retrofit scenario, the amount of energy saved, and the amount of energy consumed for the life cycle cost criterion.

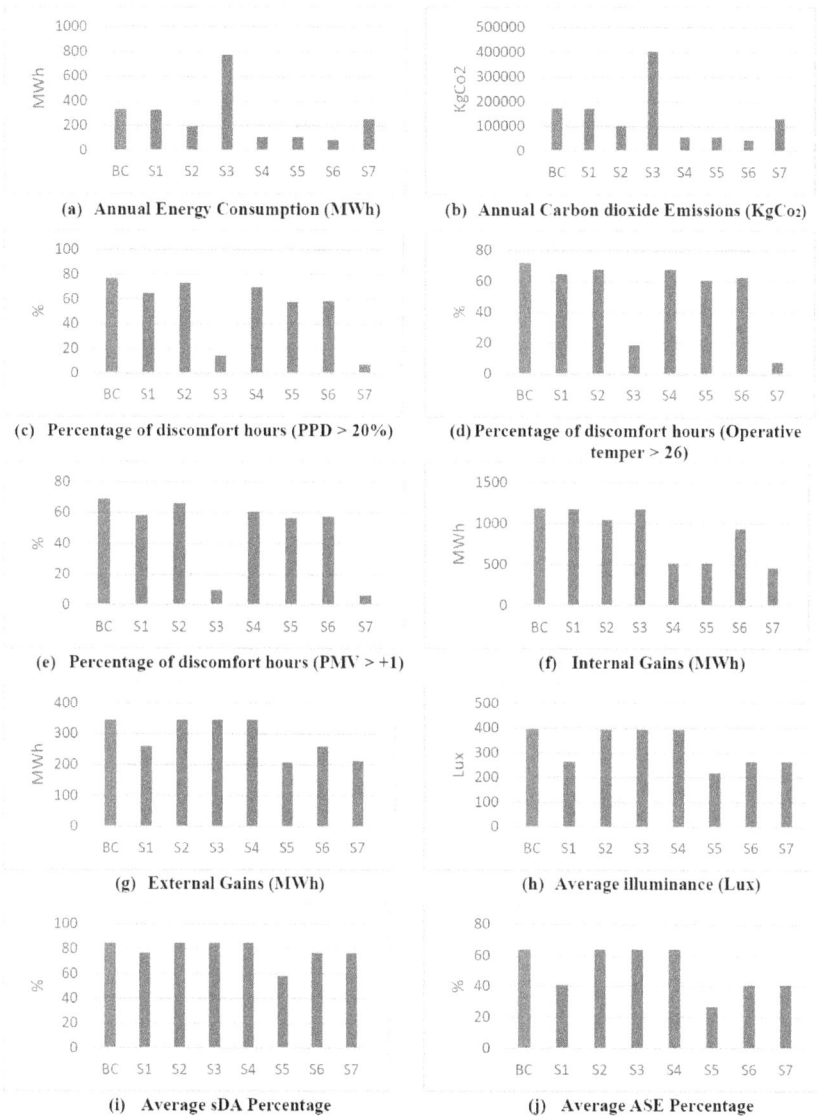

Figure 6.7 Impact of the Retrofit Scenarios on the Operational Performance; a) Annual energy consumption (MWh), b) Annual carbon dioxide emissions (KgCo₂), c) Percentage of discomfort hours (PPD > 20%), d) Percentage of discomfort hours (Operative temperature > 26 ⁰C), e) Percentage of discomfort hours (PMV > +1), f) Internal gains (MWh), g) External gains (MWh), h) Average illuminance (Lux), i) Average sDA (%), and j) Average ASE (%).

The lifetime expenses include the cost of the building's energy use throughout its lifetime (i.e., 50 years [57]) and the lifetime costs of the retrofit scenarios. On the other side, the lifetime energy savings are projected based on the energy savings resulting from the retrofit scenario over its lifetime.

Table 6.3 Main Criteria and Associated Attributes

Criteria	Attribute	Explanation
Environmental impact	*Annual energy consumption (MWh)*	Total amount of electricity used annually by the mosque's systems and operations
	Carbon dioxide emissions (KgCo2)	Total annual carbon dioxide emissions associated with the consumed energy by the mosque
Building heat gains	*Internal heat gains (MWh)*	Amount of heat gained from the building's systems and operation
	External heat gains (MWh)	Amount of heat gained from the solar radiations and the building envelope characteristics
Daylighting performance	*Illuminance (Lux)*	Average illuminance (lux) inside the space
	Spatial daylight autonomy (sDA)	Average percentage of floor area receives at least 300 lux for at least 50% of the annual occupied hours
	Annual sunlight exposure (ASE)	Average percentage of working plane area that exceeds a specified illuminance level (1,000 lux) is more than a specified number of hours (250 of the occupied hours)
Thermal comfort	*Predicted mean vote (PMV)*	The percentage of hours during the summer months from (May to October) that the PMV is greater than 1
	Percentage of people dissatisfied (PPD)	The percentage of hours during the summer months from (May to October) the PPD is greater than 20%
	Operative temperature	The percentage of hours during the summer months from (May to October) that the average operative temperature inside the space is greater than 26°

6.7.5.2   Stage 2: Using AHP Technique for Calculating the Weights of the Criteria and Attributes

By considering the defined assessment criteria and experts' judgments, the AHP method developed by Saaty [58] is used to assign relative importance to each criterion. A questionnaire was developed and distributed to experts to determine the relative importance of the criteria and their associated attributes using the scale that was generated by Saaty [59]. Using data from 26 responses, the pairwise comparison matrix and criteria weights were constructed. Each questionnaire's consistency was evaluated, and inconsistent replies were eliminated. Table 6.4 lists the criteria, attributes, and global weights. The consistency ratio (CR) is approximately 3%, which indicates that the obtained responses are sufficient because it is lower than Saaty's proposed CR requirement of 10%.

6.7.5.3 Stage 3: Ranking the Different Retrofit Scenarios to Select the Best Scenario

The TOPSIS technique, which Hwang and Yoon [60] created, ranks retrofit scenarios based on their closeness to the ideal positive solution to choose the optimum retrofit scenario among the analyzed scenarios. Table 6.5 of the initial decision matrix lists the outcomes of the retrofit scenarios' effect on the operational performance of the mosque. Moreover, the scenarios are ranked from best to worst depending on the determined performance score. Table 6.6 contains a listing of scenarios' distances, performance scores, and ranks.

To select the best retrofit scenario based on the obtained retrofit scenarios' results, the TOPSIS model is used for ranking them based on their performance score. The highest performance score is obtained in Scenario 6, which is 65.51%, because of the great enhancement to the mosque's operational performance concerning energy consumption, thermal comfort, daylighting performance, and lifetime energy savings. On the other hand, the lowest performance score is obtained by Scenario 1, which is 31.52%, where the life cycle costs were high, and the daylighting performance was low compared to the Base Case. The remaining scenarios are ranked differently based on how far they are from the ideal positive solution and how they perform. Therefore, these results show that the buildings' operational performance is not dependent only on energy consumption. Other factors, such as thermal comfort, daylighting performance, building heat gains, and life cycle costs, should be considered. According to the current condition of the mosque building and the decision-making results, Scenario 6 is considered the best retrofit scenario according to its impact on enhancing the operational performance.

Table 6.4 Summary of the Weights of the Criteria and the Global Weights

Main Criteria	Weights	Sub-Criteria	Weights	Global Weights
Environmental impact	0.131	Energy consumption	0.547	0.072
		Co_2 emissions	0.453	0.059
Thermal comfort	0.289	PMV	0.25	0.072
		PPD	0.267	0.077
		Operative temperature	0.483	0.14
Building gains	0.168	Internal gains	0.377	0.063
		External gains	0.623	0.104
Daylighting performance	0.183	sDA	0.314	0.057
		ASE	0.331	0.061
		Illuminance	0.354	0.065
Life cycle cost	0.229	Lifetime cost	0.320	0.073
		Lifetime energy savings	0.680	0.156

Table 6.5 Decision Matrix for Retrofit Scenarios and Evaluation Attributes

Scenario ID	EC	Co_2	PMV	PPD	OT	IG	EG	sDA	ASE	LUX	LCC	LES
BC	331.7	172166	68.8	76.9	72	1180	345	85	64	395.1	2,493,566.09	–
S1	331.7	172166	58.5	64.8	65	1180	260.53	76.76	40.8	263.8	4,096,298.34	–
S2	198.1	102833	66.2	73.3	68	1046	345	85	64	395.7	1,821,265.46	1,004,342.57
S3	796.39	413327	10	14.4	19	1180	345	85	64	395.1	8,450,289.26	–
S4	111.63	57939	60.8	69.9	68	518.65	345	85	64	395.1	839,182.34	1,654,383.75
S5	111.63	57939	56.6	58	61	518.65	208	58.4	26.9	220.7	4,362,164.93	1,654,383.75
S6	90	46880	57.8	58.6	63	938	260.53	76.76	40.8	263.8	2,710,964.84	1,816,988.01
S7	257	133388	6.4	7.3	7.9	464	212	76.76	40.8	263.8	8,221,828.08	561,559.80

Note: EC: Annual Energy Consumption (MWh), Co_2: Annual Carbon dioxide emissions ($KgCo_2$), PMV: Percentage of discomfort hours (PMV > +1), PPD: Percentage of discomfort hours (PPD > 20%), OT: Percentage of discomfort hours (Operative Temperature > 26 degrees), IG: Internal Heat Gains (MWh), EG: External Heat Gains (MWh), sDA: Average Spatial Daylight Autonomy (%), ASE: Average Annual Sunlight Exposure (%), LUX: Average Illuminance (Lux), LCC: Lifetime cost, LES: Lifetime Energy Savings.

Table 6.6 Ranking of Retrofit Scenarios Using TOPSIS Procedures

Scenario ID	Distance from A^+	Distance from A^-	Performance Score	Rank	$R_i(\%)$
	d_i^+	d_i^-	R_i		
BC	0.119519592	0.055409401	0.316753672	7	31.68
S1	0.113683167	0.052336091	0.315241085	8	31.52
S2	0.082669571	0.083665372	0.502993362	5	50.30
S3	0.121810269	0.062296466	0.338371469	6	33.84
S4	0.068953943	0.113044871	0.621129712	3	62.11
S5	0.061662138	0.110072065	0.640944339	2	64.09
S6	0.061499	0.116794025	0.655067831	1	65.51
S7	0.075375824	0.094513253	0.556323307	4	55.63

6.8   Conclusion

The building sector is the major contributor to the rapid increase in energy consumption and associated carbon dioxide emissions. This has led to the development of various policies and rating systems for developing energy-efficient buildings and retrofitting the existing ones according to defined assessment metrics that aid in enhancing the building's performance. But when it comes to heritage buildings, they should be treated carefully due to their unique architectural features and cultural significance. Heritage buildings are significantly important in their value. However, they are poor regarding their operational performance, as they are considered high-energy consumers compared to the other building types, have poor indoor environmental quality concerning temperature, humidity, presence of hazardous gases, as well as bad occupants' thermal comfort, unrenovated buildings have poor building envelope characteristics. This type of building has a great potential to perform efficiently when applying suitable sustainable interventions that enhance the building's operational performance. Buildings' full data should be known to determine a suitable sustainable intervention. That's why new technologies such as laser scanning, thermography, and others are important to obtain heritage-building data accurately and quickly. Also, it is important to determine the proper analysis and assessment methods for evaluating the impact of sustainable intervention on the buildings' operational performance. And at last, a multicriteria decision-making model can be used for prioritizing, ranking, and choosing the best sustainable intervention for the building based on predefined criteria.

6.9   Acknowledgment

This research was funded by the Academy of Scientific Research and Technology (ASRT), Egypt, JESOR-Development Program – Project ID: 5251.

References

[1] Al-Sakkaf, A., et al., *Sustainability assessment model for heritage buildings.* Smart and Sustainable Built Environment, 2021. (In press).

[2] Mawed, M., A. Al-Hajj, and A.A. Al-Shemery, *The Impacts of Sustainable Practices on UAE Mosques' Life Cycle Cost* in *The First International Conference of the CIB Middle East and North Africa Research Network (CIB-MENA 2014)*. 2014. Abu Dhabi, UAE.

[3] IEA, *Tracking Buildings 2021*. 2021 [20 December 2021]; Available from: https://www.iea.org/reports/tracking-buildings-2021.

[4] Lidelöw, S., et al., *Energy-efficiency measures for heritage buildings: A literature review.* Sustainable Cities and Society, 2019. **45**: p. 231–242.

[5] Cho, K.H. and S.S. Kim, *Energy performance assessment according to data acquisition levels of existing buildings.* Energies, 2019. **12**(6): p. 1149.

[6] William, M.A., et al., *Building envelopes toward energy-efficient buildings: A balanced multi-approach decision making.* International Journal of Energy Research, 2021. **45**(15): p. 21096–21113.

[7] Siksnelyte, I., et al., *An overview of multi-criteria decision-making methods in dealing with sustainable energy development issues.* Energies, 2018. **11**(10):, p. 2754.

[8] Doan, D.T., et al., *A critical comparison of green building rating systems.* Building and Environment, 2017. **123**: p. 243–260.

[9] Zhou, Z., et al., *Achieving energy efficient buildings via retrofitting of existing buildings: A case study.* Journal of Cleaner Production, 2016. **112**: p. 3605–3615.

[10] Kristl, Ž., A. TemeljotovSalaj, and A. Roumboutsos, *Sustainability and universal design aspects in heritage building refurbishment.* Facilities, 2020. **38**(9/10): p. 599–623.

[11] Al-Sakkaf, A., A. Bagchi, and T. Zayed, *Evaluating life-cycle energy costs of heritage buildings.* Buildings, 2022. **12**: p. 1271. DOI: 10.3390/buildings12081271.

[12] Conejos, S., M.Y.L. Chew, and E.H.K. Yung, *The future adaptivity of nineteenth century heritage buildings.* International Journal of Building Pathology and Adaptation, 2017. **35**(4): p. 332–347.

[13] Leissner, J., et al., *Climate for culture: Assessing the impact of climate change on the future indoor climate in historic buildings using simulations.* Heritage Science, 2015. **3**(1): p. 38.

[14] Khalil, A., S. Stravoravdis, and D. Backes, *Categorisation of building data in the digital documentation of heritage buildings.* Applied Geomatics, 2021. **13**(1): p. 29–54.

[15] Cimino, V., et al., *Interaction between daylighting and artificial lighting in relation to conservation and perception, according to new illumination system of Sistine Chapel.* Journal of Cultural Heritage, 2022. **58**: p. 256–265.

[16] Lo Faro, A. and F. Nocera, *Daylighting Design for Refurbishment of Built Heritage: A Case Study.* in *Sustainability in Energy and Buildings 2021*. 2022. (Eds. Littlewood, J.R., Howlett, R.J. and L.C. Jain), Singapore, Springer Singapore: p. 341–351.

[17] Marzouk, M., M. ElSharkawy, and A. Mahmoud, *Analysing user daylight preferences in heritage buildings using virtual reality.* Building Simulation, 2022. **15**(9): p. 1561–1576.

[18] Akande, O.K., et al., *Performance evaluation of operational energy use in refurbishment, reuse, and conservation of heritage buildings for optimum sustainability.* Frontiers of Architectural Research, 2016. **5**(3): p. 371–382.

[19] Marzouk, M. and N. Seleem, *Assessment of existing buildings performance using system dynamics technique.* Applied Energy, 2018. **211**: p. 1308–1323.

[20] Egwunatum, S., E. Joseph-Akwara, and R. Akaigwe, *Optimizing energy consumption in building designs using building information model (BIM).* Slovak Journal of Civil Engineering, 2016. **24**(3): p. 19–28.

[21] Le, A.T.H., et al., *Sustainable refurbishment for school buildings: A literature review.* International Journal of Building Pathology and Adaptation, 2021. **39**(1): p. 5–19.

[22] Stanojević, A.D., et al., *Developing multi-criteria model for the protection of built heritage from the aspect of energy retrofitting.* Energy and Buildings, 2021. **250**: p. 111285.

[23] Webb, A.L., *Energy retrofits in historic and traditional buildings: A review of problems and methods.* Renewable and Sustainable Energy Reviews, 2017. **77**: p. 748–759.

[24] Roberti, F., et al., *Energy retrofit and conservation of a historic building using multi-objective optimization and an analytic hierarchy process.* Energy and Buildings, 2017. **138**: p. 1–10.

[25] Ibrahim, H.S.S., et al., *Assessment of passive retrofitting scenarios in heritage residential buildings in hot, dry climates.* Energies, 2021. **14:** p. 3359. DOI: 10.3390/en14113359.

[26] Marzouk, M., M. ElSharkawy, and A. Eissa, *Optimizing thermal and visual efficiency using parametric configuration of skylights in heritage buildings.* Journal of Building Engineering, 2020. **31**: p. 101385.

[27] Rocha, G., et al., *A scan-to-BIM methodology applied to heritage buildings.* Heritage, 2020. **3**(1): p. 47–67.

[28] Pawłowicz, J.A., *3D modelling of historic buildings using data from a laser scanner measurements.* Journal of International Scientific Publications: Materials, Methods Technologies, 2014. **8**: p. 340–345.

[29] Zhang, Y., et al., *3D laser scanning technology-based historic building mapping for historic preservation a case study of Shang Shu Di in Fujian Province, China.* International Review for Spatial Planning and Sustainable Development, 2015. **3**(2): p. 53–67.

[30] Gardzińska, A., *Application of terrestrial laser scanning for the inventory of historical buildings on the example of measuring the elevations of the buildings in the old market square in Jarosław.* Civil and Environmental Engineering Reports, 2021. **31**(2): p. 293–309.

[31] Previtali, M., et al., *An integrated approach for threat assessment and damage identification on built heritage in climate-sensitive territories: The Albenga case study (San Clemente church).* Applied Geomatics, 2018. **10**(4): p. 485–499.

[32] Jo, Y.H. and S. Hong, *Three-dimensional digital documentation of cultural heritage site based on the convergence of terrestrial laser scanning and unmanned aerial vehicle photogrammetry.* ISPRS International Journal of Geo-Information, 2019. **8**(2): p. 53.

[33] Previtali, M., et al., *Automatic façade modelling using point cloud data for energy-efficient retrofitting.* Applied Geomatics, 2014. **6**(2): p. 95–113.

[34] Ghosh, D., et al. *Inspection of heritage structure using infrared thermography.* in *Conference and exhibition of the Indian Society for NDT (December).* 2017. *Chennai, Tamil Nadu, India.*

[35] Paoletti, D., et al., *Preventive thermographic diagnosis of historical buildings for consolidation.* Journal of Cultural Heritage, 2013. **14**(2): p. 116–121.

[36] Pereira, L.D., V. Tavares, and N. Soares, *Up-to-date challenges for the conservation, rehabilitation and energy retrofitting of higher education cultural heritage buildings.* Sustainability, 2021. **13**(4): p. 2061.

[37] Wang, C. and Y.K. Cho, *Performance evaluation of automatically generated BIM from laser scanner data for sustainability analyses.* Procedia Engineering, 2015. **118**: p. 918–925.

[38] Lyu, M., X. Zhu, and Q. Yang, *Condition assessment of heritage timber buildings in operational environments.* Journal of Civil Structural Health Monitoring, 2017. 7(4): p. 505–516.

[39] Huijbregts, Z., et al., *A proposed method to assess the damage risk of future climate change to museum objects in historic buildings.* Building and Environment, 2012. 55: p. 43–56.

[40] Karmaker, C., et al., *A framework of faculty performance evaluation: A case study in Bangladesh.* International Journal of Research in Advanced Engineering Technology, 2018. 4(3): p. 18–24.

[41] Hu, S., et al., *Environmental and energy performance assessment of buildings using scenario modelling and fuzzy analytic network process.* Applied Energy, 2019. 255: p. 113788.

[42] Kokaraki, N., et al., *Testing the reliability of deterministic multi-criteria decision-making methods using building performance simulation.* Renewable and Sustainable Energy Reviews, 2019. 112: p. 991–1007.

[43] Si, J., et al., *Assessment of building-integrated green technologies: A review and case study on applications of Multi-Criteria Decision Making (MCDM) method.* Sustainable Cities and Society, 2016. 27: p. 106–115.

[44] Dall'O', G., et al., *A multi-criteria methodology to support public administration decision making concerning sustainable energy action plans.* Energies, 2013. 6(8): p. 4308–4330.

[45] Tan, T., et al., *Combining multi-criteria decision making (MCDM) methods with building information modelling (BIM): A review.* Automation in Construction, 2021. 121: p. 103451.

[46] D'Alpaos, C. and P. Bragolusi. *Prioritization of Energy Retrofit Strategies in Public Housing: An AHP Model.* in *New Metropolitan Perspectives.* 2019. (Eds. Calabrò, F., Bevilacqua, C. and L.D. Spina), Cham: Springer International Publishing: p. 534–541.

[47] Dirutigliano, D., C. Delmastro, and S. Torabi Moghadam, *A multi-criteria application to select energy retrofit measures at the building and district scale.* Thermal Science and Engineering Progress, 2018. 6: p. 457–464.

[48] Samakie, A. *Al-Hussein Mosque.* 1979 [20 September 2020]; Available from: https://archiqoo.com/locations/al_hussein_mosque.php.

[49] ASHRAE, *ASHRAE Standard 55-2010: Thermal Environmental Conditions for Human Occupancy.* 2010. American Society of Heating, Refrigerating and Air-Conditioning Engineers.

[50] Fanger, P.O., *Thermal Comfort. Analysis and Applications in Environmental Engineering.* 1970. Copenhagen: Danish Technical Press. 244.

[51] Atmaca, A.B. and G.Z. Gedik. *Determining heat losses and heat gain through the building envelope of the mosques.* in *Proceedings of the International Research Conference on Sustainable Energy, Engineering, Materials and Environment.* 2017. Newcastle UK.

[52] Xu, Y., et al., *A multi-objective optimization method based on an adaptive meta-model for classroom design with smart electrochromic windows.* Energy, 2022. 243: p. 122777.

[53] Lee, J., M. Boubekri, and F. Liang, *Impact of building design parameters on daylighting metrics using an analysis, prediction, and optimization approach based on statistical learning technique.* Sustainability, 2019. 11(5): p. 1474.

[54] The Daylight Metrics Committee, *IES Spatial Daylight Autonomy (sDA) and Annual Sunlight Exposure (ASE). Approved Method IES LM-83-12.*2012. *Illuminating Engineering Society of North America.* New York US.

[55] El-Darwish, I.I. and R.A. El-Gendy, *The role of fenestration in promoting daylight performance. The mosques of Alexandria since the 19th century.* Alexandria Engineering Journal, 2016. **55**(4): p. 3185–3193.

[56] El Fouih, Y., et al., *Post energy audit of two mosques as a case study of intermittent occupancy buildings: Toward more sustainable mosques.* Sustainability, 2020. **12**(23): p. 10111.

[57] Rocchi, L., et al., *Sustainability evaluation of retrofitting solutions for rural buildings through life cycle approach and multi-criteria analysis.* Energy and Buildings, 2018. **173**: p. 281–290.

[58] Saaty, T.L., *A scaling method for priorities in hierarchical structures.* Journal of Mathematical Psychology, 1977. **15**(3): p. 234–281.

[59] Saaty, T.L., *Decision making with the analytic hierarchy process.* International Journal of Services Sciences, 2008. **1**(1): p. 83–98.

[60] Hwang, C.-L. and K. Yoon, *Methods for Multiple Attribute Decision Making. in Multiple Attribute Decision Making: Methods and Applications A State-of-the-Art Survey*, C.-L. Hwang and K. Yoon, Editors. 1981. Berlin, Heidelberg: Springer. p. 58–191.

7 Managing Heritage Assets using Blockchain Technology

Nouran Labib

MSc Student, Integrated Engineering Design Management Program, Faculty of Engineering, Cairo University, Postal Code 12613, Giza, Egypt

Mahmoud Metawie

Assistant Professor, Structural Engineering Department, Faculty of Engineering, Cairo University, Postal Code 12613, Giza, Egypt

Mohamed Marzouk

Professor of Construction Engineering and Management, Structural Engineering Department, Faculty of Engineering, Cairo University, Postal Code 12613, Giza, Egypt

7.1 General

The operation and maintenance of heritage buildings are critical aspects of preserving our history over the long term. Therefore, many initiatives have been recorded to save heritage buildings from environmental problems, climate change, negligence, decline, and lack of funding. The importance of preserving heritage buildings for future generations is deemed essential and requires a best-practice conservation approach. This research proposes a data-driven Chatbot framework based on Blockchain technology to help manage maintenance activities through Heritage Building Information Modeling (HBIM) technology. The proposed framework assists governments and asset managers in adopting HBIM technology to support the management and preservation of heritage buildings. Also, the proposed framework allows stakeholders to request and update asset information through trusted channels. An actual case study of Tekkeyet El Golshany heritage building in Cairo, Egypt, is presented to demonstrate the functionality of the proposed framework.

7.2 Introduction

Heritage buildings significantly impact any society because of their influence in shaping its present and future and are essential to human awareness of their history and identity over time. These types of buildings are defined according

DOI: 10.1201/9781003357483-7

to Egyptian law 144/2006 as "A building or structure with historical, symbolic, architectural, urban, or social value. It has been agreed that buildings and structures considered heritage or distinguished architectural style should be accepted by society, considered as a cultural and social phenomenon, and have the ability of stability and continuity" [1]. Despite being part of history, some of these buildings are still in use today. Some other buildings have deterioration problems, even with a broad potential for reuse. Moreover, heritage buildings face challenges in survival due to a lack of awareness, funding, and maintenance. Therefore, adopting a best practice and new technology should be applied to achieve optimum preservation and ensure that heritage buildings continue to serve their functional or cultural value to pass them on to future generations safely. There are significant shortcomings in heritage preservation. These shortcomings hinder the process of preserving heritage buildings from sustainable economic development. Some of those shortcomings are

- The lack of widely recognized and accepted methodologies for assessing cultural values, as well as difficulties in comparing the results of assessments of economic and cultural values [2].
- Lack of overall vision for management and maintenance. All Egyptian authorities dealing with cultural heritage have taken a single approach without considering the differences in these sites [3].
- The neglected impact of development projects in dealing with heritage. These impacts have brought fundamental changes to the heritage region [4].
- Reliance on local solutions and traditional treatments in urban conservation in the heritage regions of Egypt. This has led to the disappearance of some cultural values in these areas [4].
- The conservation and development of Egyptian heritage rely on international funding and UNESCO support [5].
- Uncertainties about the costs and timelines of developing heritage refurbishments and interventions due to lack of information [5].
- Many stakeholders from different backgrounds are involved in cultural heritage intervention projects. These stakeholders tend to work separately, producing scattered data, duplicating work, and ignoring existing information and input from other stakeholders [5].

This research proposes a framework for integrating work processes for documenting, recording, and managing information about heritage buildings throughout the operation and maintenance phase and presenting a communication protocol between stakeholders to enhance the data capture process and create a documentary for storing the building history. The research attempts to integrate modern technology into the heritage building conservation framework, such as building information modeling (BIM) and laser scanning, to document the history of heritage buildings in a user-friendly manner. Also, the framework aids in improving security issues in data documentation.

7.3 Literature Review

7.3.1 Building Information Modelling

BIM is a new process that is gaining popularity in architecture, engineering, and construction. It allows the creation of virtual building models, which can be linked with digital data, text, photos, and other types of information. According to the International Organization for Standardization (ISO 19650) [6], BIM is defined as "The use of a shared digital representation of a built object (including buildings, bridges, roads, process plants, etc.) to facilitate design, construction, and operation processes to form a reliable basis for decisions". This approach focuses on the concept of 3D modeling, including data for all building components that allow stakeholders to share project data throughout the project life cycle, from design to operation and maintenance phases. It is a good resource for stakeholders to make critical decisions at any project phase. The Egyptian Code for BIM [7] indicated that the complexity of the construction industry requires a new method for managing information flow throughout the project life cycle. Therefore, replacing layout fashions for layout and production within the shape of papers and 2D drawings could be more efficient. This challenge led to the formation of the so-called BIM. It allows different stakeholders to provide and share project information on a single data source throughout the project life cycle.

Implementing BIM allows a significant transformation within the construction industry, mainly in tasks that contain an excessive stage of information, unique synthetic items, a variety of stakeholders, and exaggerated pleasant designs and required outcomes. Using BIM allows different stakeholders to validate the data shared between the model and thus allows a more reliable source of information to be shared among all stakeholders throughout the project life cycle to help take corrective maintenance actions. These benefits help stakeholders work from one source. This means all data are shared among stakeholders and help reduce discrepancies between documents, such as paper documents. Therefore, it is efficient for the project cost and time.

Recent research has explored the implementation of BIM technology in historical environments, supporting the digital preservation of collected information and visualization of 3D digital models enriched with relevant information. In this research, McArthur [8] presented a BIM framework for the operation and maintenance phase of an existing building with a focus on modeling aspects and data transfer to the model. BIM Legacy is also used in data enrichment throughout the operation and maintenance process to help visualize the heritage building information. Dore and Murphy [9] also presented a unique approach to the digital recording of cultural monuments. They proposed a system using BIM software to model heritage buildings using laser scans and photographic survey data. The system uses reverse engineering, allowing the architect to model the building components separately from point cloud data and map them on one central model. Consequently, when surveying is performed by photogrammetry technology or laser scanning, the point cloud or its mesh is modeled into a BIM object.

Diakite and Zlatanova [10] presented a framework for integrating BIM and GIS. This framework allows GIS to convert BIM coordinates to actual coordinates to efficiently integrate data stored in BIM and 3D GIS environments. Portier et al. [11] presented research on archaeological archive preservation and consultation, communication functions, information exchange, and the creation of virtual museums for the old parts of the heritage city. Di Giulio et al. [12] presented the INCEPTION project, which is related to the innovation of 3D heritage modeling with an integrated approach to time-dynamic 3D reconstruction of buildings, and social environments, with 3D laser scanning for moisture detection and semiautomatic generation of heritage buildings. Cheng et al. [13] used laser scanning and photogrammetry to map the cultural heritage site in a restoration case in Taipei, Taiwan. These technologies have enabled efficient and accurate surveying of complex structures from remote locations, which was impossible with conventional surveying methods. In addition to these developments, digital information systems for displaying, analyzing, and archiving cultural heritage documents are evolving.

7.3.2 Heritage Building Information Modelling

HBIM captures and models an existing building [9]. HBIM usually includes more extensive intervention projects and highly protected buildings that require life cycle management. Researchers are currently interested in studying HBIM applications in operation and maintenance processes [14]. Although heritage buildings and sites are of great interest to government agencies, the application of HBIM in this area is insufficient [15]. Volk [16] announced that the application of HBIM in building environments is rare and requires further enhancements in data modeling, data update, and data integration. Recently, the spread of laser scanning and photogrammetry has brought the mapping of cultural heritage to the forefront. HBIM has become a well-known tool for conversations in heritage buildings. Implementing HBIM involves a reverse engineering solution in which parametric objects representing architectural components are mapped to laser scan or photogrammetric survey data [17]. This process involves multiple phases to obtain the final product. First, collect and process laser survey data, then use survey data to build a parametric model, and finally integrate the parametric model into external data sources to create technical survey drawings and documents [18].

Asset management is used to track the maintenance of heritage buildings to prevent them from deteriorating over time and keep them safe for the next generation. So, the importance of managing these buildings has been recognized over the past years [19]. Today, all studies are investigating the possibility of using automatic documentation in heritage asset management systems. Due to a lack of records and inconsistencies in data, there are challenges related to the preservation of heritage buildings and archaeological sites [20]. Moreover, there is no standard process or framework for managing heritage assets. HBIM may effectively solve this problem. However, this requires testing in the heritage context and an implementation guide for the heritage industry to facilitate adoption. BIM-enabled asset management has been viewed from a new construction perspective, but few studies have explored this possibility for heritage buildings. Some challenges impact the adoption, such as identifying essential information requirements, providing actionable guidance for including information requirements in contract documents, and the need to define information management processes [21]. These obstacles do not apply solely to implementing BIM-enabled heritage asset management [22].

7.3.3 Blockchain Applications

Blockchain is a new technology used for data exchange between untrusted parties. A lot of research has emerged to explore the applicability of Blockchain technology; among them are Stuart Haber and Scott Stornetta, who announced a framework for linking crypto chains to blocks in 1991 [23]. The purpose of this framework is to store data that cannot be deleted or modified at any time to ensure the reliability of the information. After that, Bayer presented a framework for validating transactions. The data collected from the validated transaction is stored in blocks [24]. Satoshi Nakamoto created the Blockchain network in 2008. Blocks work based on the concept of hashes that connect each block to the entire network by giving each block a hash number. A lot of companies are launching their research to develop Blockchain technology. IBM, for example, opened its research in Singapore in July 2016. In November 2016, the World Economic Forum group discussed developing a governance model for Blockchain technology. The Global Blockchain Forum introduced the Accenture Trade Group to the Digital Chamber of Commerce in 2016 [25].

There are significant shortcomings in the traditional systems of data exchange. Some of those shortcomings are [26]

- **Centralized:** Traditional systems use centralized networks. This centralized system does not provide security to the organization, as it can be attacked. Decentralized systems can eliminate or replace the traditional systems by verifying the consensus of different users in the network.
- **Trust:** Third parties can make it difficult to trust someone because they can cause significant problems in the current situation or violate data and increase redundancy. In such scenarios, Blockchain can use cryptographic hashes to create the immutable nature of blocks and ledgers, preventing failures and building trust.
- **Security:** Blockchain technology uses blocks linked with encryption keys and immutable ledgers to record transactions, making it extremely difficult for attackers to modify or delete information or transactions.
- **Transparency:** Blockchain transactions are decentralized; they are open to all users on the network and can be audited transparently.

7.3.4 Chatbot Applications

A Chatbot is a computer program that uses a user-friendly interface to mimic human interactions and view text conversations. The Chatbot model is designed to retrieve or update data depending on the user input. Chatbots can be integrated with different platforms to support the user. Many researchers tested the Chatbot's functionality; Adel et al. [27] developed a Chatbot for data collection and retrieval through textual conversations for information exchange and management systems for construction firms based on Blockchain technology. Tsai et al. [28] developed a Chatbot for construction management to automate the site equipment management processes. Kosugi and Uchida [29] developed a Chatbot that provides disaster information when disaster strikes. The user can share data about the disaster with others. Chatbot systems have been implemented in many fields. Compared to traditional delivery systems, usually built on desktop computers, Chatbot systems can provide a more efficient and immediate interface for stakeholders to query information and make good decisions.

7.4 Proposed Framework

Many technologies have been raised to preserve cultural heritage from human and natural disasters. Among the innovative technologies is artificial intelligence that, through various devices, allows obtaining data from the user. These data are processed in secured cloud computing models and Big Data architectures that get knowledge through data analysis. These results lead to achieving the balance between accurate recording and performance optimization. This chapter aims to provide a framework for exchanging heritage buildings' information through BIM throughout the operation and maintenance phase. A Blockchain model supports the framework to ensure data security between project stakeholders through a decentralized system. A Chatbot model provides ease of use for exchanging and managing maintenance information.

The proposed framework is designed based on some aspects. The first aspect is the type of Blockchain network selected as a private network to ensure the privacy of data shared among stakeholders. The second aspect is data validation to ensure that all the stakeholders validate the data stored on the Blockchain network before being documented. The third aspect is automating the storage process on the HBIM model through queries to ensure ease of data exchange between the Blockchain network and the HBIM model. The fourth aspect is the use of the Chatbot. It ensures that all stakeholders can update and manage maintenance data through a user-friendly interface. These aspects create an automated framework for managing maintenance information through a secured network in a user-friendly interface. The framework is divided into five managerial stages to fulfill the above-mentioned aspects, as illustrated in Figure 7.1.

The stakeholders include several entities such as consultants, administrative bodies (e.g., the ministry of antiques), and contractors. The system developers in administrative bodies hold the overall governance to build the Blockchain network and the Chatbot model. Further, the system developer is responsible for enrolling the actors and identifying their roles. The number of actors can be scaled anytime to fit new requirements or contractual aspects. The actors of administrative bodies are responsible for validating/updating the data broadcasted to the Blockchain network within the Communication Layer. The contractor actors update the maintenance data through the Chatbot model within the Data Exchange Layer. The consultant is responsible for retrieving the data stored on the Blockchain network through the Chatbot model within the Data Exchange Layer and using it for updating the HBIM model within the Storage Layer.

7.4.1 HBIM Model

Information about the heritage building needs to be collected to develop a model. This information can be extracted from existing reports, drawings, and survey data,

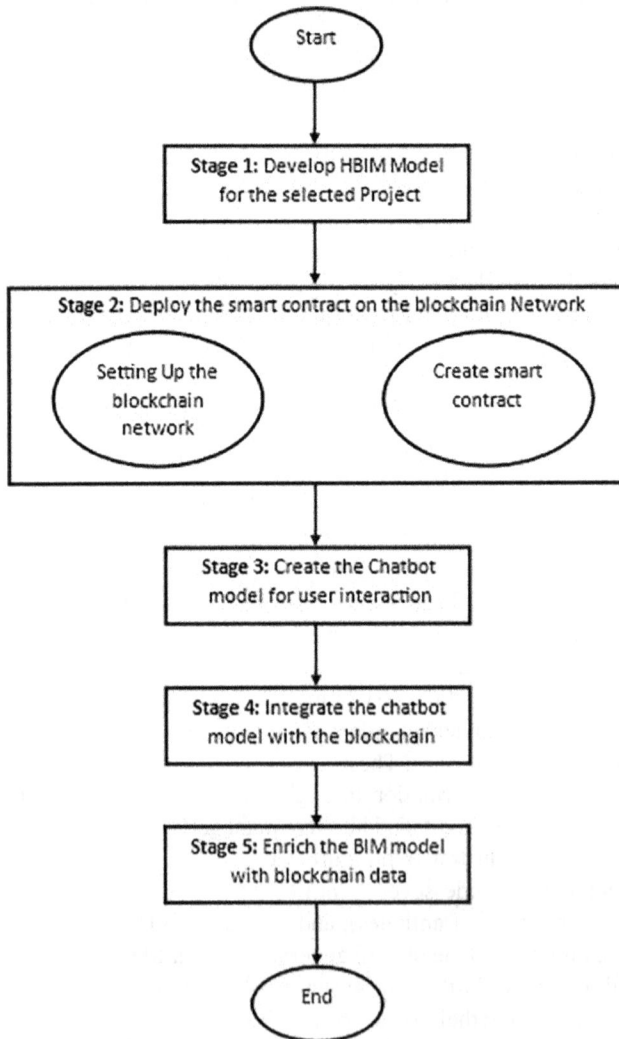

Figure 7.1 Proposed Framework Designated Stages

to create the 3D model of heritage buildings through HBIM technology. This technology allows the visualization and management of different data from different sources of information. Therefore, this research proposes a framework for managing heritage buildings using HBIM technology. This framework is divided into three phases, as shown in Figure 7.2.

7.4.2 Data Collection

According to UNESCO, Heritage documentation means "the act of acquiring, processing, presenting and recording the necessary data for the determination of the position and the actual existing form, shape, and size of a monument in the 3D space at a particular given moment in time" [30]. Heritage documentation is the primary step for any heritage preservation efforts. In Egypt, the current practice mainly depends on 2D documentation techniques, with the data being acquired with traditional survey works and photogrammetry.

The current documentation and data acquisition approach is time-consuming, labor-intensive, and prone to human error. Moreover, it is not practically adequate to capture all the fine details and imperfections of the heritage buildings. This research uses Light Detection and Ranging (LiDAR) and HBIM technologies through a flexible methodology suitable for Egyptian Heritage Documentation efforts [31]. 3D laser scanning adopts the low LiDAR technique. It releases laser beams at targeted points to measure their relative distances. The scans result in a high-density point cloud that depends on the utilized scanner. It is worth noting that 3D laser scanners can operate day or night and are almost unaffected by normal work conditions. Their use results in much faster, more accurate geometric documentation of heritage buildings than traditional techniques [32, 33]. The Data Collection process, which has been used for the investigations, was performed by the Construction Engineering Technology Lab (CETL) – Faculty of Engineering – Cairo University. CETL specializes in generating 3D documentation of heritage buildings. This process is beyond the scope of research.

7.4.3 Data Processing

The output of the 3D laser scanning process is a 3D point cloud that requires data cleaning to remove any unwanted data, such as adjacent buildings, moving traffic or persons, and atmospheric effects, as illustrated in Figure 7.3.

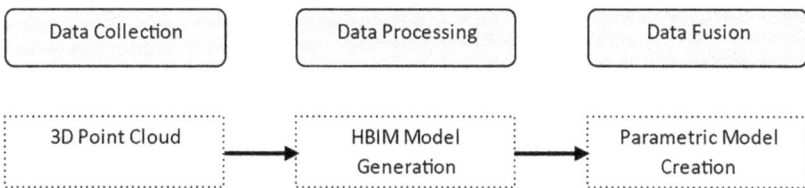

Figure 7.2 HBIM Model Creation Stages

Figure 7.3 Processed 3D Point Cloud in Autodesk Recap

The modeling of the HBIM model starts once the 3D point cloud has been processed, and it consists of the following:

- Modeling main body skeleton
- Creating families for regular structural and architectural components
- Creating families for irregular structural and architectural components
- Adding Heritage parameters to all families
- Visualizing the HBIM model

7.4.4 Data Fusion

The HBIM model is composed of different components. These components are classified based on the structural components of the building (e.g., Wall, Floor, Column, etc.). Each building component is given a unique ID to be used later in exchanging data with the external data source, as illustrated in Figure 7.4.

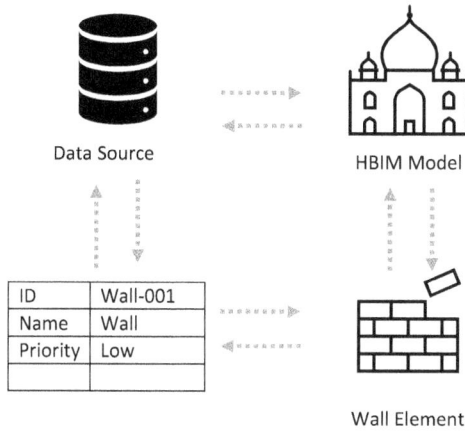

Figure 7.4 HBIM Model Data Exchange with External Data Source

Also, in this stage, Data parameters are identified. The data parameters adopt the procedure of the Historic Environment in Scotland [34], which are classified into two main categories. The first category represents parameters of activities before maintenance, and the second is parameters after maintenance activities. These data parameters are associated with components in the HBIM model using unique identifiers. These parameters can be enriched by selecting the relevant details in the model or by importing data from an external data source. Once the data are stored on the model in the form of predefined parameters, these data can be queried to train a Chatbot or stored in the model for later retrieval.

7.4.5 HBIM Model

Ad hoc maintenance is used to preserve heritage buildings. This type of maintenance is still used, where maintenance is performed just when a part of the building breaks down. Nevertheless, there is always a need for more data to study and know the history of buildings. Therefore, the quick decision to repair part of the building will take time due to the need for more documentation. Because of this challenge, it is necessary to create a framework for managing heritage buildings by recording the data of heritage buildings in chronological order. These data are used for routine maintenance and quick decisions in critical cases. This chapter aims to illustrate the required data to create an HBIM model to assist stakeholders through the operation and maintenance phase.

Relying on traditional documentation methods such as historical records and existing drawings can be time-consuming and labor-intensive, and the data collected can be inaccurate. Therefore, LiDAR, digital photogrammetry, and Territorial Land Scanning (TLS), as some of the most recent technologies applied for surveying and mapping, are gaining efficiency in heritage, archaeological, and landscape studies. It is practical and robust, especially in data collection, site surveys, and modeling processes. An appropriate surveying method should be selected according to the specific characteristics and requirements of the goals and tasks.

Based on the above discussion, this research focuses on the heritage building of Tekkeyet El Golshany and introduces TLS and point cloud modeling as the main methodologies.

7.5 Chatbot Model Development

The proposed Chatbot model is developed using IBM Watson Assistant [35]. This model is used as a communication medium for updating and retrieving maintenance information of heritage buildings, such as building components maintenance procedures, open maintenance tasks, maintenance location, maintenance life photos, and scheduled maintenance activities in a structured and automated platform during the operation and maintenance phase. This model in heritage management will address challenges such as proactive messaging, scalability, user-friendliness, decision support, and data storage. The system consists of a Chatbot and a backend model that connects to the Instant Messaging (IM) application. The IM application receives the user's text input and sends it to the Chatbot. The Chatbot model then analyzes the user's message and uses several features to generate a response, as shown in Figure 7.5. The back-end system contains a database to store the extracted information obtained from the HBIM model. This information is automatically retrieved at the user's request. The conversation begins with the Chatbot recognizing a message representing the user's intent to request or update the data. Subsequent conversation flows are designed so that the Chatbot requires input from the user. The conversation ends when the user updates the requested information.

The IBM Watson Assistant dialog consists of nodes. Each node can contain one or more intents. This intent identifies the user intent from the chatting with the chatbot. The user is then asked a predefined question to capture or update the user with the required information in entities. Table 7.1 lists the conversation flow based on user input to the Chatbot.

Figure 7.5 Proposed Chatbot Model Workflow

Table 7.1 The Flow of Conversation in the Chatbot

User input: "Query All Data."
The service recognizes the user's goal in the input and maps it to an intent.
Intent: #Queryall_Transactions
You define an answer to the question about the data needed by adding a dialog node for it.
Dialog node: #Queryall_Transactions
Response: Success [data retrieval]

7.6 Blockchain Model Development

The proposed Blockchain model is developed using the IBM Blockchain Platform [36]. The model is used to manage and validate heritage building information and maintenance data, such as maintenance procedures for building components, open maintenance tasks, maintenance locations, maintenance lifetime photos, and planned maintenance activities on a secure and reliable platform among all stakeholders. By employing this model in heritage management, some challenges preventing the industry from using HBIM will be addressed, such as confidentiality, provenance tracking, disintermediation, nonrepudiation, multiparty aggregation, recordkeeping, change tracing, and data ownership. Creating a digital building information model allows people to interact with the building to optimize their actions.

The model, combined with Blockchain technology, develops trust among various stakeholders, which aids in reducing contractual issues such as construction disputes and supply chain failures. The platform leverages the development of building information modeling to manage 3D models that support the preservation of cultural heritage information [37]. The platform contains 3D models that support the storage of cultural heritage information. Also, it allows 3D models to be saved and shared with associated details, enabling external users to update and improve the stored knowledge. The search engines and comprehensive web interfaces provide flexible and fast access to the platform. The platform is considered the central historical database repository for preserving and sharing HBIM models, semantic information, documentation, and all other possible contributions external users make. Also, it provides a trusted channel for stakeholders to manage and verify the data entered by the user through a predefined contract term (Smart Contracts).

The smart contract is a program stored on the Blockchain that is typically used to automate the execution of contracts so that all involved parties can see the results immediately, without the involvement of intermediaries or the loss of time. The workflow can also be automated to trigger the following action when the conditions are met [38]. Smart contracts work based on conditions written in code on the Blockchain. The computer's network acts when the specified conditions are met and validated. Once the transaction is completed, the Blockchain is updated. The transaction cannot be modified, and only authorized parties can view the results [39].

The Blockchain consortium is used in this research to enable the exchange of information among all stakeholders within a private channel. The consortium's Blockchain provides privacy and prevents unregistered individuals from participating in transaction process validation. The transaction flow of the Blockchain network is divided into four processes: transaction proposal, proposal implementation, transaction validation, and transaction notification. The transaction proposal is based on the Blockchain network from the client application with the data required by the user. The responsible peer validates the data, and the signature returns the response to the proposal. Therefore, confirmation from all relevant stakeholders can significantly improve transaction reliability. Once a transaction is created, it is sent to all responsible peers in the channel for data validation. A new block is built on the Blockchain network when the transaction is validated. Once created, this block will be sent to all peers in the Blockchain network channel. Then, the client is notified that a new transaction has been made. This transaction is completed when validated by the responsible peer. This transaction also forms a new block that will be added to the Blockchain. Peers are notified about the creation of new blocks. If the responsible peer has not validated the transaction, it will remain in the transaction queue and not be added to the Blockchain network.

Building Information Exchange tools exchange information from external databases extracted from the Chatbot model into the HBIM database. The building information exchange tool provides a functional HBIM template to integrate the external databases with the project parameters within the HBIM model. The parameters that enrich the model can be customized to include helpful information for operation and maintenance. One of the challenges of heritage protection is that the components, materials, and structures of each building are different from other buildings. This uniqueness hinders standardizing the parameters used to normalize the data exchange methods of heritage buildings.

Furthermore, the research focuses on data collection and modeling maintenance data requirements rather than tackling the BIM process. The user can also update this template automatically using data queries in the database to be imported to the HBIM model for documentation purposes. In BIM fields, strategy and procedure definitions need to be traced back to the asset data for a particular model, that is, the parameters and information that are integrated with the model. In addition, the building life cycle can be managed by applying BIM procedures to incorporate information into the model. This integration can be done by linking an external data source to the model using various queries in the model database. Autodesk Revit uses the Microsoft Access database to connect HBIM information to external sources.

7.7 Framework Implementation

A case study of Tekkeyet El Golshany heritage building is presented to demonstrate the framework's applicability. Tekkeyet El Golshany heritage site is in Fatimid Cairo, Egypt, and was built in the early 16th century. The site comprises 12 private rooms, a kitchen, a prayer hall, and a mausoleum. The main central building is the mausoleum is nearly 15 m in height, topped by a dome. The cladding of the main façade is ceramic tiles, as illustrated in Figure 7.6. All that remains today is the mausoleum and four rooms in disrepair.

The building geometry has been scanned using 3D laser scanning by the Construction Engineering Technology Lab, Faculty of Engineering, Cairo University. The scanning process documents building elements and passes them to future generations. The data obtained from the scanning process is a 3D point cloud used as a reference for creating the building on Autodesk Revit. To record the state of the

Figure 7.6 Tekkeyet El Golshany Heritage Building Façade

building, components used to record the actual state of the building were investigated. Building components are also grouped into more management activities that help assess the condition of the building as a whole. As shown in Figure 7.7, management components are given a unique ID, later used for data enrichment.

Building components are determined based on Scotland's historic environment and linked to maintenance activities through project parameters that fall into two categories: premaintenance activities and post-maintenance activities. These project parameters are identified in the Autodesk Revit model. The data to enrich the project parameters must contain keywords for later use to train the Chatbot, as shown in Table 7.2. The information stored about the parameters is extracted into database files containing various tables. Relationships between these tables are also exported. After that, the HBIM template is then created in Microsoft Excel format. It contains parameters for building components associated with unique IDs for later updating model data. This template is linked to the HBIM model database as a query, as shown in Figure 7.8, to update the database with the latest project information. The HBIM template is updated through the data retrieved from user interaction with the Chatbot, as shown in Figure 7.9.

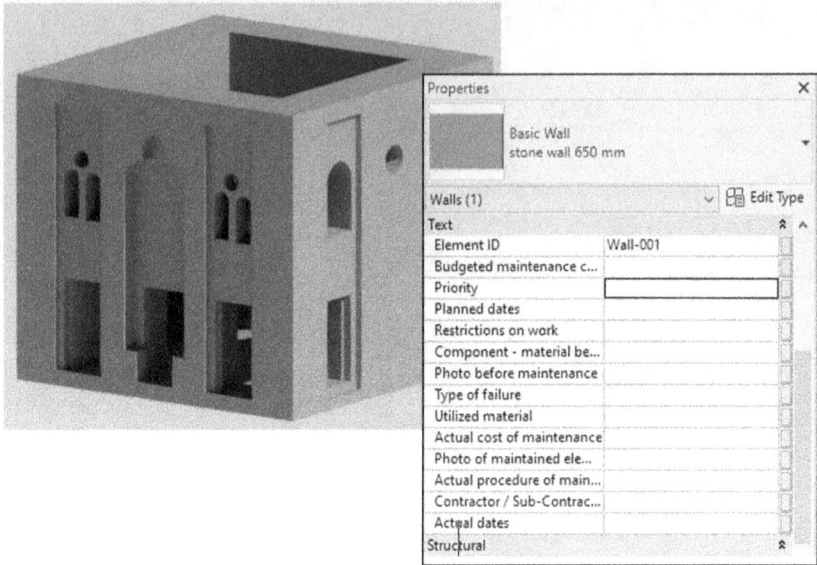

Figure 7.7 Wall Unique ID

Table 7.2 Project Parameters Keywords

Project Parameters	Assumptions
Actual cost of repair	Actual Cost
Actual date	dd/mm/yyyy
Actual procedure of repair	Text
Budgeted maintenance cost	Budget Cost
Component of material before maintenance	Concrete, Block, Stone, Brick
Contractor	Arab Contractors, Orascom
Material supplier	Motaheda, El-Slab, Future
Photo before repair	https://1drv.ms/u/s!ArbPCGCOowmY300aYEfDl xV9jD3K?e=oteLF5
Photo of the repaired component	https://1drv.ms/u/s!ArbPCGCOowmY300aYEfDl xV9jD3K?e=oteLF5
Planned date	dd/mm/yyyy
Priority	Low, Medium, High
Proposed procedure for repair	Paint windows, replace the window, wall furnish
Restrictions on work	No shifts
Type of failure	Cracks
Utilized material	Concrete, Block, Stone, Brick

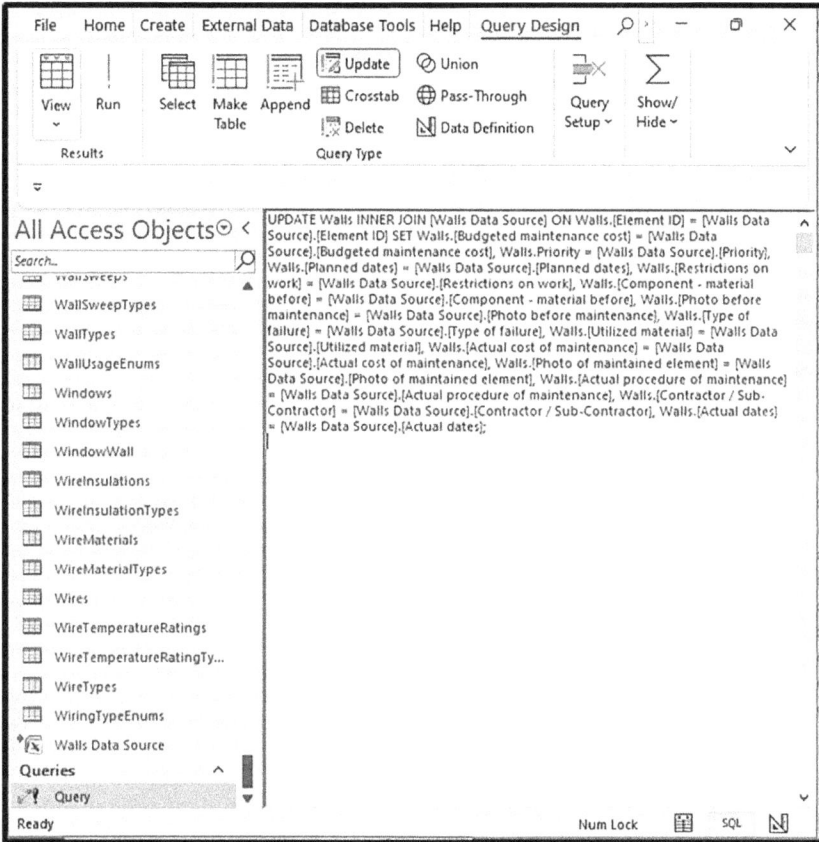

Figure 7.8 Microsoft Access Data Queries

The data entered by the user to the Chatbot is temporarily stored as variable parameters in IBM Watson Assistant, as shown in Figure 7.10. Then, the variable parameters are passed through the webhook to the IBM function, as shown in Figure 7.11.

These data are then sent to the smart contract uploaded on the IBM Blockchain network, as shown in Figure 7.12. Once these data are sent to the Blockchain network, a new transaction is created. This transaction is verified and saved as a new block on the Blockchain network. This Blockchain network is broadcast to all peers in the Blockchain channel. When the user requests the data stored on the network, the data will be retrieved in JSON format, as shown in Figure 7.13. These data are then used to update the HBIM model.

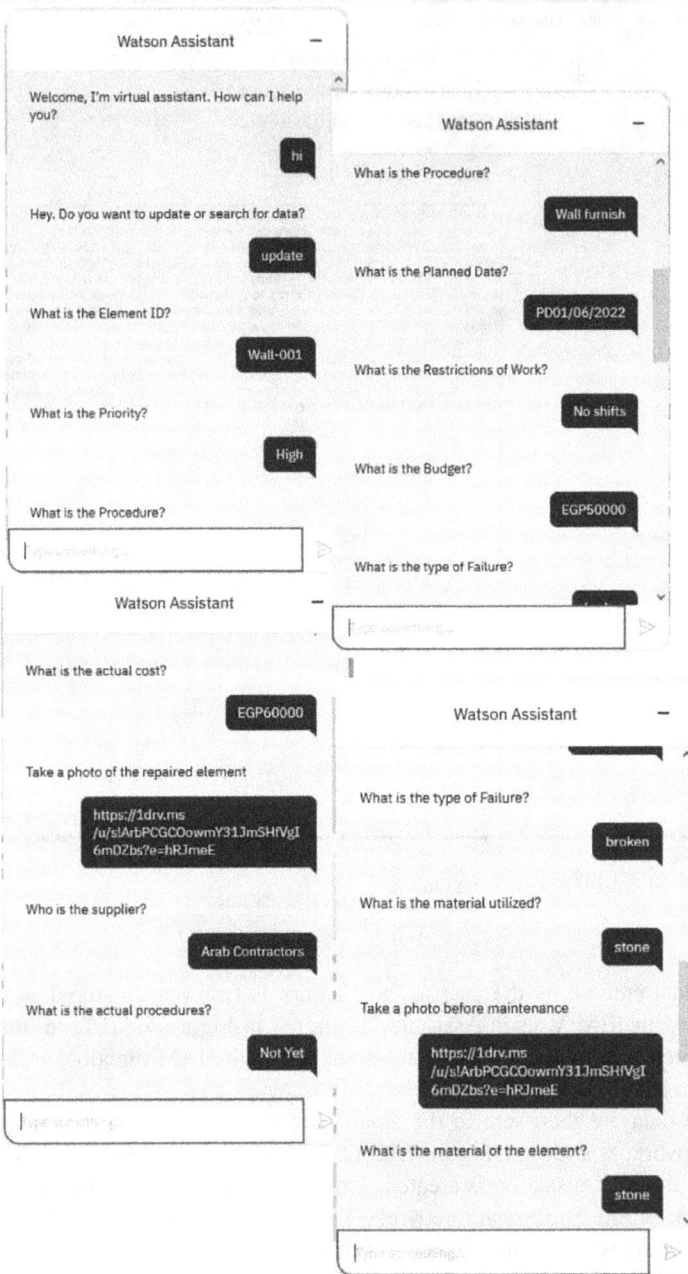

Figure 7.9 User Updating Building Data

	Check for	Save it as	If not present, ask	Type		
1	@Update	$fcn	fcn?	Required	⚙	🗑
2	@elementId	$pwId	What is the Elem	Required	⚙	🗑
3	@Priority	$P1	What is the Prior	Required	⚙	🗑
4	@Proposedproced	$P2	What is the Proc	Required	⚙	🗑

Update_Tranasctions   Customize ⚙   ×

Node name will be shown to customers for disambiguation so use something descriptive.   Settings

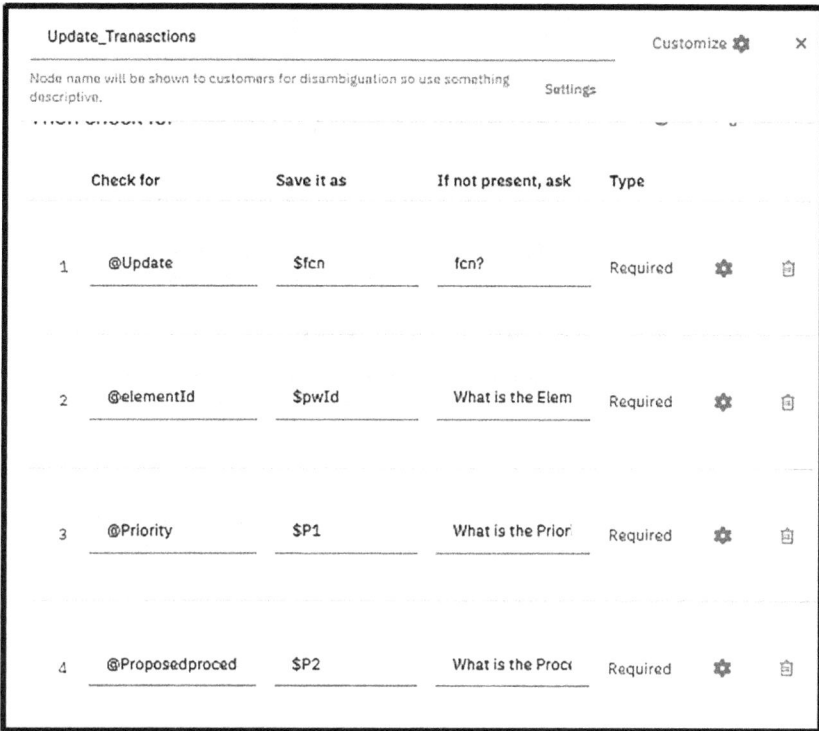

Figure 7.10 Updated Maintenance Information saved as Variable Parameters

Then call out to my webhook

Learn more

Parameters

Key	Value
P1	"$P1"
P2	"$P2"
P3	"$P3"
P4	"$P4"
P5	"$P5"

Figure 7.11 Call Out Webhook to Pass Data to IBM Functions

```
class PwContract extends Contract {

    async pwExists(ctx, pwId) {
        const buffer = await ctx.stub.getState(pwId);
        return (!!buffer && buffer.length > 0);
    }

    async createPw(ctx, pwId, P1, P2, P3, P4, P5, P6, P7, P8, P9, P10, P11, P12, P13) {
        const exists = await this.pwExists(ctx, pwId);
        if (exists) {
            throw new Error(`The pw ${pwId} already exists`);
        }
        const asset = { P1, P2, P3, P4, P5, P6, P7, P8, P9, P10, P11, P12, P13 };
        const buffer = Buffer.from(JSON.stringify(asset));
        await ctx.stub.putState(pwId, buffer);
    }

    async readPw(ctx, pwId) {
        const exists = await this.pwExists(ctx, pwId);
        if (!exists) {
            throw new Error(`The pw ${pwId} does not exist`);
        }
```

Figure 7.12 Parameters used to Capture User Data in Smart Contract

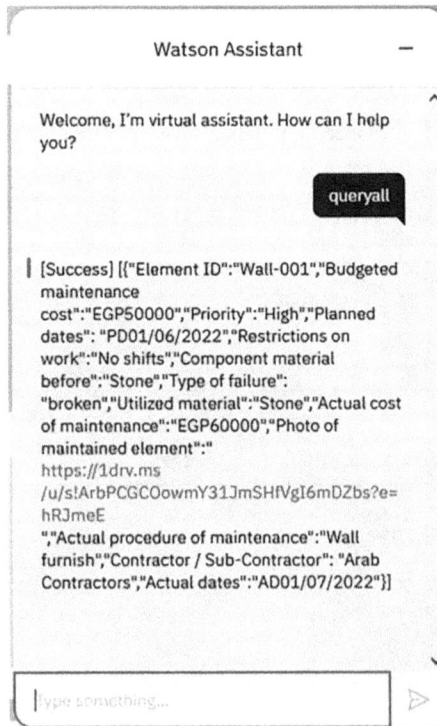

Figure 7.13 User Request Data

7.8 Conclusion

The proposed framework helps users extract information about maintenance activities from the HBIM model and return the updated data via a data-driven Chatbot. Information found through conversations between users and Chatbot is passed to the Blockchain network and validated by stakeholders before being stored in the HBIM model. Integrating HBIM and the Blockchain environment improves the management of transactions between the Blockchain model and the HBIM model. In addition, this framework enhances Asset Information Management (AIM) capabilities by storing digital project data. Therefore, by using accurate data, asset owners can improve decision-making, optimize the use of operating assets, avoid culture and language misunderstandings among stakeholders, and eliminate waste in the asset process. Blockchain also helps owners, consultants, contractors, and suppliers upload and visualize data anytime, anywhere, throughout the life cycle of a heritage building. A case study was presented to highlight the proposed framework's features and capabilities.

References

[1] H. Shalaby, "VVITA project strategies of conserving heritage buildings in egypt," December, 2021.

[2] B. Derman, "Cultures of development and indigenous knowledge: The erosion of traditional boundaries," *Africa Today*, vol. 50, no. 2, pp. 67–85, 2003, doi: 10.2979/aft.2003.50.2.66.

[3] I. Serageldin, *Cultural Heritage and Development in the Arab World*, 2008, [Online]. Available: http://archnet.org/library/documents/one-document.jsp?document_id=13021

[4] T. Fushiya, "Archaeological site management and local involvement: A case study from abu rawash, Egypt," *Conservation and Management of Archaeological Sites*, vol. 12, no. 4, pp. 324–355, 2010, doi: 10.1179/175355213x13789818050224.

[5] Y. Yilmaz and R. El-Gamil, "Cultural heritage management in Turkey and Egypt: A comparative study," *Advances in Hospitality and Tourism Research*, vol. 6, no. 1, pp. 68–91, 2018, doi: 10.30519/ahtr.446254.

[6] C. Castainghristoph, "From a project management perspective Table of Contents," *EFCA BIM and ISO 19650 from a Project Management Perspective*, vol. 1, no. 1, 2018, [Online]. Available: https://www.frinet.dk/media/1307/381783_efca_flipbook_bim_fri.pdf

[7] The Egyptian Code for Building Information Modeling (BIM), "The egyptian code for Building Information Modeling (BIM)," pp. 1–48, 2018.

[8] J. J. McArthur, "A Building Information Management (BIM) framework and supporting case study for existing building operations, maintenance and sustainability," *Procedia Engineering*, vol. 118, pp. 1104–1111, 2015, doi: 10.1016/j.proeng.2015.08.450.

[9] C. Dore and M. Murphy, "Current state of the art historic building information modelling," *International Archives of the Photogrammetry, Remote Sensing and Spatial Information Sciences – ISPRS Archives*, vol. 42, no. 2W5, pp. 185–192, 2017, doi: 10.5194/isprs-archives-XLII-2-W5-185-2017.

[10] A. A. Diakite and S. Zlatanova, "Automatic geo-referencing of BIM in GIS environments using building footprints," *Computers Environment and Urban Systems*, vol. 80, p. 101453, 2020, doi: 10.1016/j.compenvurbsys.2019.101453.

[11] S. Potier, J. L. Maltret, and J. Zoller, "Computer graphics: Assistance for archaelogical hypotheses," *Automation in Construction*, vol. 9, no. 1, pp. 117–128, 2000, doi: 10.1016/S0926-5805(99)00054-0.

[12] R. di Giulio, F. Maietti, E. Piaia, M. Medici, F. Ferrari, and B. Turillazzi, "Integrated data capturing requirements for 3D semantic modelling of cultural heritage: The inception protocol," *International Archives of the Photogrammetry, Remote Sensing and Spatial Information Sciences – ISPRS Archives*, vol. 42, no. 2W3, pp. 251–257, 2017, doi: 10.5194/isprs-archives-XLII-2-W3-251-2017.

[13] H. M. Cheng, W. bin Yang, and Y. N. Yen, "BIM applied in historical building documentation and refurbishing," *International Archives of the Photogrammetry, Remote Sensing and Spatial Information Sciences – ISPRS Archives*, vol. 40, no. 5W7, pp. 85–90, 2015, doi: 10.5194/isprsarchives-XL-5-W7-85-2015.

[14] X. Xiong, A. Adan, B. Akinci, and D. Huber, "Automatic creation of semantically rich 3D building models from laser scanner data," *Automation in Construction*, vol. 31, pp. 325–337, 2013, doi: 10.1016/j.autcon.2012.10.006.

[15] D. Ilter and E. Ergen, *BIM for building refurbishment and maintenance: Current status and research directions*, vol. 33, no. 3, 2015, doi: 10.1108/SS-02-2015-0008.

[16] R. Volk, J. Stengel, and F. Schultmann, "Building Information Modeling (BIM) for existing buildings – Literature review and future needs," *Automation in Construction*, vol. 38, pp. 109–127, 2014, doi: 10.1016/j.autcon.2013.10.023.

[17] M. Murphy, E. McGovern, and S. Pavia, "Historic Building Information Modelling – Adding intelligence to laser and image based surveys of European classical architecture," *ISPRS Journal of Photogrammetry and Remote Sensing*, vol. 76, pp. 89–102, 2013, doi: 10.1016/j.isprsjprs.2012.11.006.

[18] W. S. Jeong, S. Chang, J. W. Son, and J. S. Yi, "BIM-integrated construction operation simulation for just-in-time production management," *Sustainability (Switzerland)*, vol. 8, no. 11, pp. 1–25, 2016, doi: 10.3390/su8111106.

[19] K. Clark, "Power of place – Heritage policy at the start of the new millennium," *Historic Environment: Policy and Practice*, vol. 10, no. 3–4, pp. 255–281, 2019, doi: 10.1080/17567505.2019.1696549.

[20] Spanish Cultural Heritage Institute, "National Plan for Abbeys, Monasteries and Convents National Plan for Abbeys, Monasteries and Convents," Madrid: Spanish Cultural Heritage Institute, 2011.

[21] M. Kassem, N. Iqbal, G. Kelly, S. Lockley, and N. Dawood, "Building information modelling: Protocols for collaborative design processes," *Journal of Information Technology in Construction*, vol. 19, pp. 126–149, 2014.

[22] M. Murphy, E. Mcgovern, and S. Pavia, "Historic building information modelling (HBIM)," *Structural Survey*, vol. 27, no. 4, pp. 311–327, 2009, doi: 10.1108/02630800910985108.

[23] M. Modani, S. Patidar, and S. Verma, "A methodological review on applications of blockchain technology and its limitations," *Information Management and Computer Science*, vol. 4, no. 1, pp. 01–05, 2021, doi: 10.26480/imcs.01.2021.01.05.

[24] U. Bodkhe *et al.*, "Blockchain for Industry 4.0: A comprehensive review," *IEEE Access*, vol. 8, pp. 79764–79800, 2020, doi: 10.1109/ACCESS.2020.2988579.

[25] P. Tasatanattakool and C. Techapanupreeda, "Blockchain: Challenges and applications," In *2018 International Conference on Information Networking (ICOIN), Chiang Mai, Thailand*, pp. 473–475, 2018, doi: 10.1109/ICOIN.2018.8343163.

[26] S. Nanayakkara, M. N. N. Rodrigo, S. Perera, G. T. Weerasuriya, and A. A. Hijazi, "A methodology for selection of a Blockchain platform to develop an enterprise system,"

Journal of Industrial Information Integration, vol. 23, no. 04, p. 100215, 2021, doi: 10.1016/j.jii.2021.100215.

[27] K. Adel, A. Elhakeem, and M. Marzouk, "Chatbot for construction firms using scalable blockchain network," *Automation in Construction*, vol. 141, no. 4, p. 104390, 2022, doi: 10.1016/j.autcon.2022.104390.

[28] M. H. Tsai, C. H. Yang, C. H. Wang, I. T. Yang, and S. C. Kang, "SEMA: A site equipment management assistant for construction management," *KSCE Journal of Civil Engineering*, vol. 26, no. 3, pp. 1144–1162, 2022, doi: 10.1007/s12205-021-0972-2.

[29] M. Kosugi and U. Osamu, "Chatbot application for sharing disaster-information," *2019 International Conference on Information and Communication Technologies for Disaster Management (ICT-DM)*, Paris, France, 2019.

[30] O. C. Amans, Y. Y. Ziggah, and A. O. Daniel, "The need for 3D laser scanning documentation for select nigeria cultural heritage sites," *European Scientific Journal*, vol. 9, no. 24, pp. 75–91, 2013, [Online]. Available: http://eujournal.org/index.php/esj/article/viewFile/1696/1685

[31] S. van Riel, "Exploring the use of 3D GIS as an analytical tool in archaeological excavation practice," 2016, doi: 10.13140/RG.2.1.4738.2643.

[32] M. Metawie, M. Ali, and M. Marzouk, "Framework for HBIM applications in egyptian heritage building information modelling," *BUE ACE1 – Sustainable Vital Technologies in Engineering & Informatics*, pp. 1–7, 2016.

[33] J. Morales, V. Plaza-Leiva, A. Mandow, J. A. Gomez-Ruiz, J. Serón, and A. García-Cerezo, "Analysis of 3D scan measurement distribution with application to a multi-beam lidar on a rotating platform," *Sensors (Switzerland)*, vol. 18, no. 2, 2018, doi: 10.3390/s18020395.

[34] Historic Scotland, "Short guide 1: Fabric improvements for energy efficiency in traditional buildings," *Historic Scotland Short Guides*, vol. 1, p. 5, 2010.

[35] IBM Watson Assistant, "Watson Assistant: Build better virtual agents, powered by AI." https://www.ibm.com/products/watson-assistant

[36] IBM Blockchain, "IBM blockchain." https://www.ibm.com/blockchain

[37] D. Kifokeris and C. Koch, "Blockchain in construction logistics: State-of-art, constructability, and the advent of a new digital business model in Sweden," *Proceedings of the 2019 European Conference on Computing in Construction*, Chania, Crete, Greece, vol. 1, pp. 332–340, 2019, doi: 10.35490/ec3.2019.163.

[38] Z. Zheng *et al.*, "An overview on smart contracts: Challenges, advances and platforms," *Future Generation Computer Systems*, vol. 105, pp. 475–491, 2020, doi: 10.1016/j.future.2019.12.019.

[39] H. Hasan, E. AlHadhrami, A. AlDhaheri, K. Salah, and R. Jayaraman, "Smart contract-based approach for efficient shipment management," *Computers & Industrial Engineering*, vol. 136, pp. 149–159, 2019, doi: 10.1016/j.cie.2019.07.022.

Index

For Product Safety Concerns and Information please contact our EU
representative GPSR@taylorandfrancis.com
Taylor & Francis Verlag GmbH, Kaufingerstraße 24, 80331 München, Germany

9 781032 413174